American Design Ethic

W9-ADU-383

A History of Industrial Design to 1940

Arthur J. Pulos

This book was set in VIP Helvetica by Village Typographers,
Inc., and printed and bound by Halliday Lithograph in the
United States of America.

Library of Congress Cataloging in Publication Data

Pulos, Arthur J.
    The American design ethic.

    Includes bibliography and index.
    1. Design, Industrial — United States — History.
I. Title.
TS23.P84 1983   745.2'0973   82–4625
ISBN 0–262–16085–4

Contents

Preface

Design is the indispensable leavening of the American way of life. It emerged with the need of the colonists to transform the wilderness into a secure haven and expanded as a natural component of the industrial revolution in the New World. The United States was in all likelihood the first nation to be designed—to come into being as a deliberate consequence of the actions of men who recognized a problem and resolved it with the greatest benefit to the whole. America did not just happen; it was designed.

Whatever their national origins, the first settlers showed the courage, the energy, and the ingenuity that would enable America to transcend aristocratic and theocratic rule. These emigrants to the New World were more firmly held together by a concept than those who remained behind were united by religion and monarchy. Their passion for freedom of individual opportunity allowed enterprise and industry to flourish in the service of an expanding common market.

Despite the immediacy of the need to secure a new life on a strange continent, these transplanted Europeans were unable to escape the bittersweet memory of their national origins. And each succeeding band of immigrants carried along its native expression, stirring it into the common cauldron, so that even today the American culture is a roiling brew of transplanted elegance and folk ethic. However, with the mass production that followed the industrial revolution, the Americans have been able to generate a unique cultural contribution. Products designed for industry and commerce, vehicles and vessels for transportation, and mechanical and electronic appliances for the home reveal an American passion for energy-conserving devices that are at times exquisitely suited to their purposes. At first the forms of these products were determined primarily by technological factors and mechanical functions. However, as the virtuosity of productive means has been refined, the manufacturers of machine-made products have discovered a conscience that is now putting increasing emphasis on the primacy of the human operator. The products of tomorrow will make it possible for Americans to move into a postmaterialistic society in which products will serve humbly as elements of closed-loop environmental systems.

American design is constructed from the building blocks of the puritan ethic and sheathed by liberal mercantilism. It is erected upon an economy of means and a respect for natural forces, and it is illuminated with the spirit of self-reliance. It displays a faith in empirical discovery capped with theory. It presumes that there is an inescapable correlation between the perfection of a solution to a problem and the elegance of the form it takes. There is no principle of American design so powerful as that an object acquires beauty as it approaches the ideal typeform of its species.

History will prove that, if a humane democracy is to be this country's legacy to mankind, its unique contribution to world culture may well be the democracy of its manufactured products. Despite their transitory value, they are the true artifacts of the United States because in them civilizations to come will find an expressive crystallization of our life energy and our daily existence.

Acknowledgment

Note on Citations

The author's work was supported by a grant from the National Endowment for the Arts.

In citations of sources—for example, (25, 37)—the first number indicates the source's position in the bibliography and the second indicates the page in the source. An intervening number specifies a volume.

American Design Ethic

The Colonies

The Arts of Survival

. . . whence came all these people? They are a mixture of English, Scottish, Irish, French, Dutch, Germans and Swedes. From this promiscuous breed, that race now called Americans have arisen. . . .There is room for everybody in America.

J. Hector St. Joan de Crevecoeur, 1782 (25, 37)

The mother countries of the United States never imagined that their colonies would dream of independence and aspire one day to shape their own destiny. Nor was it conceivable to them that the scattered settlements in the wilderness of the New World would be fused into a new nation by the fire of a revolution. England, France, and Spain (the great European powers of the time), as well as Holland, Sweden, Germany, and other countries, looked upon the distant continent as a virgin source of wealth in the form of rare metals and minerals and forest and animal resources, as a place to establish plantations and manufactories to supplement those of the homeland, as a prospective market for the surpluses of their own industrial revolutions, and as a depository for their social and religious dissidents.

It seems inevitable that the tide of hereditary aristocracy and secular and religious autocracy that crested in Europe in the seventeenth and eighteenth centuries should have washed the seeds of independence onto the shores of North America. There were two great waves of colonialism. The wave to the south came first to seek out and bring back the treasures of the new world; the northern one that quickly followed was one of self-exiled egalitarians seeking to plant a purer society to the glory of God. For several decades, emigrants from one country or another competed for their own parts of North America. However, by the mid-1600s the English colonists and the language and manners of their mother country prevailed.

The first London Company was chartered by the Crown in 1607 to explore and exploit the resources of Virginia and to return to its shareholders a fifth of all of the riches found or mined. The colonizing band included adventurers, soldiers of fortune, unprincipled young men,

and bankrupts who were expected to trade with the Indians and to establish plantations and manufactures that would profit themselves and their sponsors. The colony that the Company founded at Jamestown was eventually led by Captain John Smith, soldier and adventurer, whose report was to be described later by Alexis de Tocqueville as breathing "that ardour for discovery, that spirit of enterprise which characterized the men of his time, when the manners of chivalry were united to zeal for commerce, and made subservient to the acquisition of wealth." (86, II, 408)

The memoirs of Thomas Jefferson shed additional light on the preoccupation with personal gain of the Englishmen who colonized the South: "At the time of the first settlement of the English in Virginia, when land was to be had for little or nothing, some provident persons having obtained large grants of it, and being desirous of maintaining the splendour of their families, entailed their property upon their descendants. The transmission of these estates from generation to generation, to men who bore the same name, had the effect of raising up a distinct class of families, who, possessing by law the privilege of perpetuating their wealth, formed by these means a sort of patrician order, distinguished by the grandeur and luxury of their establishments." (86, II, 415)

The second ship dispatched to the Virginia colony in 1608 brought over from London people skilled in manufactures. Some were put to work in the forests to produce clapboards, wainscoting, and other wood derivatives such as tar and pitch that were in short supply at home; others attempted to smelt iron ore; others erected a "glass-house" to make glass from beach sand. Under constant threat from the Indians, such efforts were eventually abandoned. Of all of the ventures undertaken in Virginia, only the plantations that were established in the warm humid climate along the eastern seaboard showed promise of success. The colonists' attempts to raise textile fibers and establish silk culture as they were pressed to do by their London shareholders were unsuccessful, but they found that tobacco was admirably suited to the climate and commanded an excellent price in European markets.

The success of the patricians was aided in no small measure by the introduction of slavery in 1620, when the first twenty Africans were put ashore on the banks of the James River. (It is an interesting coincidence that in that same year the first Puritans were landing to the north at Plymouth in search of individual freedom.) The combination of available labor and successful crops (first tobacco; later cotton, rice, and indigo) soon created an agrarian aristocracy in the South, with England as its focus. The southern colonists sent to England for clothing and for home furnishings, and shipped their children to England to be educated and indoctrinated with British political and social philosophy. The wealthy southerners came to view themselves as aristocrats whose position obligated them to develop an environment in the New World that was as elegant and fashionable as that of England. From their estates were to come the proud new Americans, jealous of their right to economic rewards, zealous in their freedom, and knowledgeable enough of history and law to realize that a universal tide was running in favor of republican equality. It was Jefferson, a southern patrician, who transposed John Locke's phrase "life, liberty, and the pursuit of property" into "life, liberty, and the pursuit of happiness," which he considered more egalitarian and therefore more acceptable to his northern compatriots. By the end of the colonial era the southerners were to realize that, although they had no tenant class, they were really one people with the freemen of the North.

The northern wave of English emigrants was made up of those disenfranchised merchants and disenchanted scholars and ecclesiastics who had dared to speak out against the practices and authority of the Anglican church. Those who called themselves Puritans questioned the influence of Roman Catholic ritual and dogma in England and insisted on their right to interpret the word of God for individual guidance. These Separatists, as they were known, were against any political appointments to church office and

called for a complete break between church and state. Queen Elizabeth resented any question of her authority over the church and used her influence and that of her court in an attempt to bring the Separatists into line by denying them access to the established and respected professions. As a result, the dissidents were forced to follow the practical theology of Calvin by turning for economic support to the scientific, industrial, and mercantile professions that were beyond conventional political and ecclesiastic control. Nevertheless, the persecution continued. The first band of Pilgrims (as they called themselves) sailed in 1609 from northern England for Holland. They remained in exile in Leyden until 1620, when, expressing a desire to preserve their English heritage, which was in danger of assimilation in Holland, they were granted a charter by the Crown to emigrate to North America. William Brewster, William Bradford, and Edward Winslow led the first band of about 150 Pilgrims on the long voyage to America. Most were less than 40 years old at the time.

The Pilgrims came to the New World sharing a philosophy that was austere and a practical ethic that would not permit indulgence in luxury and the fine arts. Their first building was described as a storehouse with six cannons on the roof and a lower part that was used as a church on Sundays. Although this first band of Pilgrims and others that soon followed had both secular and ecclesiastic differences, they were bound together by a common language and had been subject to the same laws and parallel acts of suppression. They shared a passionate desire for political and religious freedom, and they brought with them a sense of community rule that they had shared under the English parish system, a new morality born out of the Reformation, and a respect for intelligent inquiry for the gathering and refinement of useful knowledge. Individuals generally did not consider themselves superior to others.

By 1635 more than 20,000 émigrés had reached New England. Most were not Separatists, but rather Puritans who sought not so much to leave the Anglican church as to change it from within in keeping with the principles of the Reformation. However, their distance from the mother church eventually led to a complete independence, and by 1650 many of the Anglican churches in America had become Congregational. For the most part, England was not displeased by the departure of her religious dissidents. In fact, she did what she could to encourage many of them to leave by offering them a greater degree of independence than did many other countries that were seeking to establish a foothold in America. This stimulated the development of settlements that were stronger internally and led to governmental innovations. Whereas other countries ruled their colonies with an iron hand through a Crown-appointed governor, the English colonies were established either by land grants (as in the case of Pennsylvania, Maryland, New Jersey, and the Carolinas) that permitted the owner to sell land and govern as he wished under the benign eye of the Crown, or else by permitting the emigrants to form a political society under the protection of the mother country in order to govern themselves in any way that was not in contradiction with her interest.

In effect, the flow of government in the colonies was the inverse of that in England. It ran from the individual to the community, from the community to the colony, and from there to the state. By the mid-1600s every township in New England was autonomous, and though each accepted the supremacy of England the way to democracy was clear. The colonists had produced a new form of government, without precedent.

In addition to their new-found virtue of freedom, the English colonists of the North, who were primarily a middle class that had been excised from the center of an ancient feudal society, also made a virtue of intelligence and knowledge. They were well educated and had come to the New World determined to establish and nourish an intellectual community. They taxed themselves to found schools and colleges to provide educational opportunities for their children that could only have been acquired by inheritance in England. The general expansion of knowledge that followed proved that devotion to the acquisition

of knowledge could result in gains that were equal to those available through aristocratic dominance over others. "Diligent Researchers at home," testified J. E. Edwards in support of the Puritans' commitment to knowledge, "and Travel into remote Countries have produced New Observations and Remarks, unheard of Discoveries and Inventions." (30)

The Puritans' self-inflicted repressions strengthened them to face the rigors of pioneer life. *The Christian Directory of Rich and Baxter,* published in London in 1665, describes the practical ethics of the Puritans and insists that their devotion to a strenuous life benefited technological development and production. The directory is replete with homilies on industry, thrift, and spiritual salvation: "The public welfare, or the good of many, is to be valued above our own. Every man is therefore bound to do all the good he can to others, especially for the Church and the Commonwealth: And this is to be done not by Idleness, but by labor! . . . Labor not to be rich: The meaning is, that you make not Riches your chief end: Riches for our fleshly ends must not ultimately be intended or sought. But in subordination to higher things they may; That is, you may labor in the manner as tendeth most to your success and lawful gain: . . . . You may labor to be Rich for God, though not for the flesh and sin." (7) These ideas would be echoed less than a century later by Benjamin Franklin in *Poor Richard's Almanac.*

The Puritans' social principles helped to shape the northern mind and wound up the dynamic drive that still moves American enterprise. The homes of the northern merchants and businessmen were soon as full of luxuries as those of their agrarian friends in the South. As a result, the two waves of colonialism came into economic as well as political balance. In the end, both were to subscribe to a spirit of equality based on a rationalism in which service to man is taken as the proper path to personal comfort and security.

The new American was a synthesis of all of the struggles for intellectual and spiritual freedom and economic opportunity that had been contained for centuries beneath the hardened and polished surfaces of the established social and economic structures of the Old World. This prototype citizen of the New World was trusting but not gullible, honest but shrewd, friendly yet cautious, and as quick-witted as he was deliberate. He may have been poor, but he was no longer fettered to land and landlord. He was proud of the courage that had set him upon this distant land and jealous of the freedom that he had come to claim. And, even though the glowing embers of familiar habits and ancient customs might be fanned into flame whenever he paused to rest or reflect, the harsh realities of the struggle to survive in an environment that was as hostile as it was abundant taught him to respect and depend upon his own physical and intellectual powers. In contrast to the exorbitant consumption of the energy of those workmen in Europe who labored in penury and peonage in the service of political and religious aristocracies, the colonial freeman had limitless riches laid out before him if he only had the ingenuity and energy to claim them.

The rigors of the frontier did not encourage the dissipation of energy that was not materially rewarding. It is understandable that some people romanticize the early frontier settlers as dependent upon humble handicrafts, suggesting that they were unaware of the technology of the seventeenth century. However, there was no great gap in availability of implements and machines between the rural craftsman and the urban artisan. Both knew and used tools and machines to supplement human and animal muscle with the natural forces of wind and water.

Away from the more sophisticated trades of the colonial towns, the pioneer American learned quickly that survival in the wilderness depended upon knowledge of those trades that were essential to frontier living, such as hunting, fishing, farming, lumbering, animal husbandry, blacksmithing, and carpentry. He was obliged to clear and plant his own land, to construct a shelter, to fabricate and repair his own implements, and to make harnesses for his animals and shoes for himself and his family. He sheared wool and grew flax so that the womenfolk could spin and weave linsey-woolsey and other homespun fabrics for clothing and linens. Every farmer was his own

mechanic, and every home was a manufactory in which adults and children worked together to serve their own needs. And a family that was more diligent or knowledgeable in one home industry or another could barter or sell its surplus products to others.

Later, Alexander Hamilton described the nation as "a vast scene of household manufactures," and Tocqueville recognized versatility as a special and even unique quality of the colonials. Although he acknowledged that "this circumstance is prejudicial to the excellence of the work," Tocqueville said that it "powerfully contributes to awaken the intelligence of the workman" and that "if the American be less perfect in each craft than the European, at least there is scarcely any trade with which he is utterly unacquainted." (86, I, 510) Thus, the new American found his security ensured not by a lifetime commitment to a single craft or skill, but by a versatility that reduced the danger of occupational enslavement. His experiences along the frontier stimulated the qualities of self-reliance and faith in independent action.

The frontier settlers realized quickly that courage and an independent spirit would not in themselves ensure success. Although the implements brought from Europe may have been appropriate for the smaller plots of land and the limited woods of England and the Continent, they proved less than adequate in the American wilderness. Driven by the press of survival and the promise of natural riches to be had for the taking, the colonists were obliged to adapt their tools and weapons.

The first implements to show the influence of the new environment were the weapons that were essential for defense and the hunt, the implements that were necessary for clearing the land and working the soil, and the tools needed to build shelters, prepare food, and make clothing. When an implement turned out to be inadequate, was broken in use, or simply wore out, it was replaced with a new one patterned after the original. Every time this was done, the replacement was modified in one way or another in response to experience so that it could serve better with

The challenge of survival in the New World, including the clearing of the wilderness and the building of a home, encouraged inventiveness and demanded versatility. Collection of New York Public Library.

Every frontier home was a manufactory. Women spun wool on the great wheel and wound it on the reel, or spun flax to be woven into linen cloth. The wooden mortar was used to pound corn into the meal that took the place of wheat flour. Harper's Monthly, *January 1877.*

improved comfort, more efficiency and speed, or less of a physical demand. This led to an understanding, based upon personal experience, of the way function can be improved by the refinement of form.

The stubby tree-felling axe of Europe was one of the first tools to be altered by the American woodsman. He increased the weight of the poll (the blunt end of the head) for better balance so that the axe could be swung more freely and accurately, and in some cases he replaced the poll with a second cutting edge that enabled him to work twice as long in the forest without having to stop to resharpen. He also lengthened and reshaped and curved the handle so that he could handle the increased weight with greater power, comfort, and efficiency. The American axe became renowned for its functional perfection.

The short, stiff scythe of Europe was modified by the American farmer so that he could stand straighter in the field and work longer without tiring. He lengthened and curved the blade so that he could reach and hold more stalks of grain for cutting. The American scythe became a graceful implement that complemented the farmer's natural sweeping motion and enabled him to cut a cleaner and wider swath. In the eighteenth century a cradle was added so that the cut grain could be laid aside in neat rows to be gathered more easily with less loss.

By the mid-1700s, German gunsmiths in central Pennsylvania had created a rifle with an unusually long barrel that, with a ball wrapped in a greased patch, provided the accuracy that was needed to bring game down at long range in the dense forest. This strange stick of a weapon obsoleted the shorter, smooth-bore muskets of the day that were fired more or less randomly at a target in the hope that they would hit something, or if necessary frighten it away. The new weapon traveled westward with the frontiersmen across the Appalachians to become the famed Kentucky rifle that secured the way for the settlers and contributed substantially to the cause of the colonists in the War of Independence.

The axe, the scythe, and the rifle are only three of the many products that were refined by the colonists. From the very beginning of colonization, the Americans were practicing euthenics, the development of human well-being and efficient functioning through the improvement of the environment. Design—the contemporary embodiment of this principle—is concerned with the humane quality of the man-made environment. It further assumes that there is a direct correlation between that quality and the elegance of the form that it takes. Benjamin Franklin had anticipated the principle of functionalism as the relation of beauty to utility when he wrote to Charles Wilson Peale that "the invention of a machine or the improvement of an implement is of more importance than a masterpiece of Raphael. . . . nothing is good or beautiful but in the measure that it is useful. . . ." (77, 3) With that simple expression Franklin captured the essence of the American design ethic that has persisted now for more than two centuries. Franklin was undoubtedly familiar with the observation of his correspondent William Hogarth in the essay "Analysis of Beauty" (London, 1754) that "in nature's machines how wonderfully do we see beauty and use go hand in hand." No principle of design in America is so strong as that an object achieves beauty to the same degree to which it serves its function. By this philosophy it may be assumed that every product is constantly striving to achieve the perfect and therefore beautiful typeform of its species.

The principle of beauty as the natural by-product of functional refinement was given additional meaning by the spartan circumstances of the colonial environment. In the wilderness and the new settlements of America, the colonist was obliged to avoid the devotion to rich detail and elaborate ornament that served in the older aristocracies to exaggerate the value of products through the extravagant consumption of energy. Therefore, whatever esthetic reward the American was to derive from his products had to be found in the economy of means and the purification of form to purpose and from the soundness of proportion and the clarity of symbolic form that inevitably result. The natural texture of his materials and the honest marks of his tools

Frontiersmen modified the European felling axe for more efficient cutting in the dense forests of North America. Courtesy of Essex Institute, Salem, Massachusetts.

The short-handled broad axe was shaped with an offset blade and an angled handle to enable the builder to hand-hew logs into square timbers for construction. Index of American Design, National Gallery of Art.

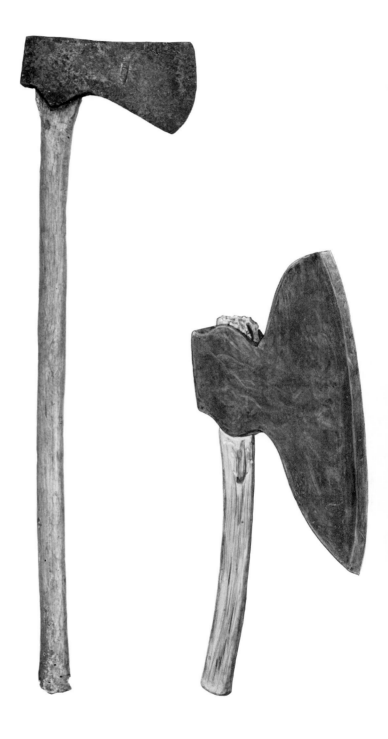

The European scythe was modified until it was admirably suited to natural body movements. In the next century, a cradle would be added to increase efficiency. Catalog of Dinkson-Davidson Hardware Company.

The Kentucky rifle owed its origin to the hunting gun brought to Pennsylvania by German immigrants. Its long barrel proved more accurate for game hunting in the dense forests of America. T. M. Stine Collection, Smithsonian Institution.

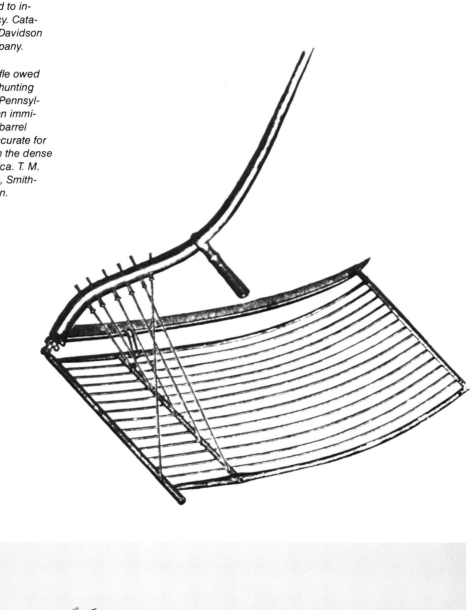

were ornament enough for the settler. The fingerprint in the clay, the scar of the adze on the wooden beam, the facets of the hammer on metal, and the warp and weft of the loom all provided the democratizing link that bound maker and consumer to product. This preference for "natural" textures as the "honest" surfaces of metal, wood, stone, brick, and fabric is still preferred to arbitrary patterning in America. Such other decorative elements as may appear from time to time carried a deeper symbolic meaning beyond the visual entertainment that they may provide today.

It is understandable that much of the technology of colonial America should have been based on wood. There was a natural relationship between the vastness, variety, and versatility of wood and the amount and quality of energy that the colonist needed to work this abundant material. The soft woods such as pine, fir, and cedar provided builders with a light yet stiff material that was readily shaped into construction materials and assembled into public buildings and private dwellings. The lean-tos of the English Pilgrims, the log cabins of the Swedish immigrants, and most of the later structures of the colonies were built with wood. Wood was often used as a substitute for masonry; for example, the German immigrants transformed the stone barns of their native land into the magnificent wooden barns that still grace the Pennsylvania countryside. Cabinetmakers and joiners built furniture that made excellent use of the strength, color, and grain of hard woods such as maple, cherry, and walnut. Tough and sinewy woods such as ash, hickory, and oak were found to be perfect for lighter furniture, farm implements, and carts. The Conestoga wagons of Pennsylvania, the schooners adapted to American winds and waters, and the "topsail" vessels for ocean travel would never have been built in such quantities in the colonies were it not for the availability of wood.

Still, some Americans were concerned over the tendency to make everything out of wood. Thomas Jefferson deplored the "unhappy prejudice" of the time that stone and masonry homes were unhealthy to live in as a convenient philosophy that reflected the cheapness of wood and the availability of carpenters. Jefferson wrote of wooden homes that "it is impossible to devise things more ugly, uncomfortable and, happily, more perishable." He believed that buildings made of permanent materials added permanent value to the state, whereas those built of perishable wood did not. It is interesting to speculate whether the American concept of obsolescence as evidence of progress may not have been related to the colonists' dependence upon wood as the primary material for dwellings and objects.

The spare utilitarian objects and the unsophisticated folk arts of the rural colonists were often demeaned by the patricians of America and England as "country made" and therefore devoid of aesthetic value. The fact, however, is that a democratic art form was beginning to emerge in the reserve of colonists' buildings and furnishings, the simple elegance of their implements, and the direct innocence of their signs and symbols.

The first stream of American design consciousness was rooted in such empirical and unpretentious adaptations to the exigencies of life along the frontiers of the New World. Moreover, as the colonists became disenchanted with the economic and political behavior of the British it became increasingly patriotic for them to depend upon native ideas and expressions of humbler origin. Later the New England Transcendentalists would affirm the principle that the true American expression must come from within rather than be garbed in borrowed raiment.

While Jean Jacques Rousseau (whose philosophy was well known to colonial intellectuals) and other European social philosophers of the time were formulating theories based on the premise that only a return to the natural man and a simpler way of life would rid civilization of its distortions and release anew those primary virtues upon which a good and sound society depends, their concepts were being put to the test instinctively in the distant wilderness and settlements. While the theorists were warning that the arts had become decadent in the service of despotism and pleading that only a return to the simpler forms of nature would rid civilization of

In functional form, the
Conestoga wagon from
Pennsylvania was the
eighteenth-century equiv-
alent of modern trailer
homes. Index of American
Design, National Gallery
of Art.

its distortions, the foundation for a simpler egalitarian society was being laid in North America. The new Americans had already begun to exhibit their unique ability to react empirically to the exigencies of their environment, and were not unwilling to leave it to others to formulate theories to explain their action.

Another social philosopher of the time, Johann Herder, proposed that the fine arts must spring by purification from the popular arts. This belief that man-made forms must be respected according to their devotion to human needs before they may be revered as aesthetic expression is entirely consistent with the attitude of Americans toward the objects that serve them: that the daily arts and folk expression must retain their naiveté, that any attempt to transform them into fine art destroys the eloquent fragility that gave them their value in the first place, and that therefore they can have value only to the degree to which they remain valid in source, pure in design, and unadulterated by mass production.

If we once grant the principle of the division of labour, then it follows that one man can live only by finding out what other men want.

Arnold Toynbee, 1884 (87, 56)

There is a persistent tendency to romanticize the colonial era as a period when unique objects were fabricated by humble craftsmen employing quaint methods rooted in antiquity. It is more accurate to note that the colonial artisans were as determined to put the most recent technological discoveries to practical use as they were to adopt the latest styles for their products. They were generally aware that the knowledge acquired from the emerging sciences, if applied to their craft, could increase their capacity for production and allow time to improve their methods of work in order to produce better products more economically and to search for those new products to manufacture that promised a greater return on their investment of energy and resources.

When Sir Francis Bacon stated in 1620 that the object of knowledge was to change the shape of man's world, he gave form to the central theme of the industrial revolution. Within two years Bacon's followers had founded the Royal Society of London, "to promote the welfare of the Arts and Sciences," and had established a distinction between the acquisition and ordering of knowledge and the application of that knowledge to industry. The impact of that revolution was not lost on the colonies, despite British generalizations that American technology depended but little upon the sober reasoning of science. As early as 1690, private evening schools had been established in New York to teach the apprentices for whom the masters were obligated to provide an elemental education. These schools multiplied as an increasing artisan population sought additional learning, not only in reading, writing, and arithmetic but also in geometry, trigonometry, and many of the more specialized trades.

Most of the master craftsmen of the time were literate and were avid readers of general newspapers and pamphlets. They also sought to further themselves by reading on such subjects as architecture and building, cabinetmaking, ironwork, and the other useful arts. Those who could not afford to build libraries of their own became members of subscription libraries such as those organized by the American Philosophical Society in Philadelphia and the Baltimore Mechanics Company. In contrast to the closely guarded mysteries of the European guilds, the American artisans realized that if they shared knowledge freely with one another the general state of industry would be advanced.

Even though the practical application of knowledge was essential for their success, the colonial artisans and gentlemen derived particular gratification from tinkering with machines and other ingenious devices. The challenge of making the forces of nature work for him awakened the restless spirit of the designer, who is forever dissatisfied with things as they are and driven to make them better. In the process, the colonial artisan was transformed into a "mechanick" preoccupied with the methods and processes by whose improvement his energy might be expanded, his security guaranteed, and his prosperity ensured. Moreover, by assigning an increasing percentage of the work through machines to lesser employees, he found time to devote himself to more learning, to experiment with new concepts for products, and to improve his merchandising capabilities.

The gentleman "mechanick" also realized that, if he could apply his knowledge of the sciences and his financial resources to the support and development of manufactures to satisfy human needs and promote comfort and happiness, he could expect a high economic return. One outstanding organization, established in Philadelphia in 1750 for this purpose, was the American Society for Promoting and Propagating Useful Knowledge. Through it and other organizations like it, churchmen, artisans, and wealthy laymen pooled their resources and combined their education, experience, and sense of product potential to refine processes and establish manufactories.

Shipbuilding companies were formed to build sloops, privateers, schooners, and, in particular, "topsail" ships for transatlantic trade. Although at first most of their ancillary parts such as cordage, sails, and metal fittings were imported from England, gradually all of these came to be made in the colonies. By the time of the Revolution almost a third of the British merchant ships were American-built, as were the majority of the ships owned and sailed by the Americans in competition with the British.

The craft of coachmaking serves as an excellent barometer of colonial affluence. The financial and human resources necessary to bring together the artisans in wood and metal needed to construct a carriage and the skilled upholsterers, leatherworkers, painters, and decorators needed to finish it only became available after the middle of the eighteenth century. In short order, however, Americans were building coaches, chariots, landaus, phaetons, post-chaises, curricles, chairs, sedans, and sleighs of all types. Since they could be sold at a lower price than the imported products, they captured a market that had been the special reserve of English coachmakers.

The promise of a ready market for ingenious devices encouraged the invention of many products for manufacture. Surveying instruments and mariner's compasses were developed by such men as Isaac Doolittle, a New Haven clockmaker. Benjamin Gale was awarded a gold medal by the Royal Society of London for a seeder, drawn by oxen, that could open a furrow, deposit seed and manure, and close it up again in one operation. And Benjamin Thompson, the English loyalist later to become Count Rumford of Bavaria, invented the modern fireplace, the drip coffeepot, and the kitchen range. Thomas Jefferson was fascinated by ingenious products and developed several for his home at Monticello, including a seven-day clock, simultaneously acting double doors, dumbwaiters, and a swivel chair.

Benjamin Franklin stands out as America's first scientist. While most of his countrymen were preoccupied with acquiring knowledge from abroad, Franklin was generating his own by conducting the definitive experiments with electricity that would bring him international fame. One of his most popular inventions was the "Pennsylvania fireplace" (now known as the Franklin

Benjamin Thompson's experiments with heat and convection currents resulted in the "drip" or percolator method of making coffee. Sanborn C. Brown, Benjamin Thompson, Count Rumford *(Cambridge: MIT Press, 1979).*

Thomas Jefferson's restless imagination led him to develop this swivel chair by modifying a Windsor. His interest in convenience presaged the contemporary designer's commitment to comfort. American Philosophical Society.

stove), which was manufactured to his specifications by Robert Grace at Warwick Furnace in Chester County. The primary significance of this invention is that it pulled the fireplace out of the chimney, where it had been part of the architecture, and treated it as a portable and therefore marketable appliance. In addition, it employed scientific principles to control the fire and to direct the heated air in a manner that increased fuel efficiency. When the Pennsylvania fireplace was first offered for sale in the *Pennsylvania Gazette* on December 3, 1741 and promoted with a brochure that might have been America's first promotional flyer, Governor Thomas offered to give Franklin a colonial patent to guarantee his profit from the invention for a number of years. However, Franklin declined it "from a Principle that has ever weigh'd with me on such Occasions, viz. That as we enjoy great Advantages from the Inventions of others, we should be glad of an Opportunity to serve others by any Invention of ours, and this we should do freely and generously." (56, 419) In a short time, others were manufacturing stoves, as they still do today, according to the design that Franklin refused to patent.

At first glance, it would seem that the expanding interest in devices that saved labor and made life more pleasant was in contradiction to the Puritan ethic that glorified labor almost as an end in itself. In fact, there still seems to be an undercurrent of guilt in the American mind at the easy life that has been made possible by technology. However, toward the end of the colonial era such reservations were set aside in the fervent desire of the colonists to establish a self-sustaining economy in the face of British restrictions. This challenge in the name of patriotism provided the industrial momentum that helped the colonies sever their political ties to England.

At first England was proud of the success of her colonies. Sir Joseph Child, a director and governor of the East India Company, praised the emigrants as a people "whose frugality, industry and temperance, and the happiness of whose laws and institutions, promise to them a long life and a wonderful increase of people, riches and power." However, within a century the success

of the colonists was beginning to attract to America talent and intelligence that could not expand or find free expression in the Old World. In response to such threats, England began to place increasingly severe restrictions on the trade and manufacturing practices of her colonies. Some products were restricted in the number that could be made, and the manufacture of other products was forbidden entirely. Even though England was following the conventional practice of the times in using trade restrictions to protect her home industries, the colonists found such limitations intolerable because, in the midst of great resources and opportunities, they were being treated as second-class citizens and forced to restrict their enterprise in order to protect a fading homeland. It is not surprising, therefore, that the "true cause" of the Revolution would be put forth as having been "not so much that the colonists were denied representation in the central government, or that they were unduly restrained in respect to any liberty of their persons, but rather that their rights to property were continually interfered with, that they were denied the privilege of freely buying and selling wherever and whenever they might see fit, and of following the occupations which seemed to them the most remunerative." (142, 708)

Although the British Navigation Acts were passed in 1650 to restrict competition from Dutch shipbuilders, before the end of the 1600s they were being used to restrict colonial shipbuilding through a ruling restricting transport between the colonies and England to British ships. Later the Navigation Acts were expanded to rule that the colonies must buy only from England those manufactured products England had to sell, and soon they were further strengthened to prevent the colonies from manufacturing any products that were made in England. In addition, by 1696 the colonial governments were being required to report the annual state of their industries to the English Board of Trade in an attempt to divert them from industrial activity. By 1708 the Board of Trade was being warned that the colonists were manufacturing most of the products that they needed, and that if some effective way was not found to stop it, they would carry it further,

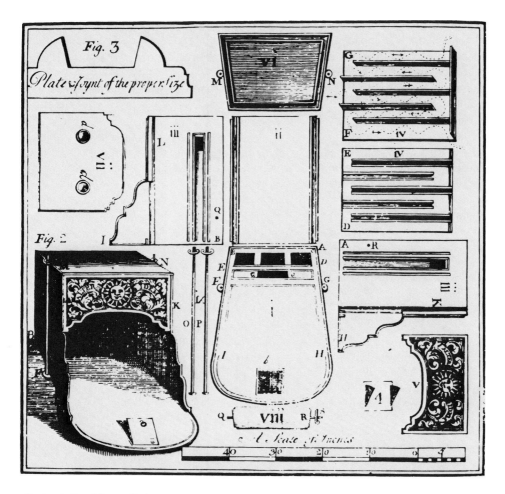

Benjamin Franklin applied
the principle of convection
to the development of a
cast-iron stove, conceived
in the form of a labyrinth
through which cooler air
was drawn to be heated
and sent out to warm the
room. Reference 56.

much to the disadvantage of English manufacturers. Soon some colonies were refusing to make reports, or else reporting a lack of any settled manufacturing. Colonists who did not consider their activities either economically or politically immoral reacted to the restrictive legislation by going underground. It has been estimated that nine-tenths of all colonial merchants and one-fourth of the signers of the Declaration of Independence were involved in contraband trade. At the very moment of the battle of Lexington, John Hancock, the "Prince of Contraband Traders," was in Boston on trial, with John Adams as his counsel, on charges based on his activities as a smuggler.

The English reacted strongly to what they considered to be unlawful and provocative acts of the colonies. General Thomas Gage summed up the push of the Americans toward economic independence in a letter to Lord Barrington in 1772, advising him that the English must "cramp their Trade as far as it can be done prudentially," and that "cities flourish and increase by intensive Trade, Artisans and Mechanicks of all sorts are drawn thither, who Teach all sorts of Handicraft Work before unknown in the Country, and they soon come to make for themselves what they used to import." (17, 616)

When the first Continental Congress met in Philadelphia on September 5, 1774, its objectives were commercial rather than political. It was agreed that after 1774 no more products were to be imported from the British Empire and that after September 1775 no products were to be exported to England. A century later, John Leander Bishop acclaimed the courage and wisdom of this position: ". . . the prohibition of their manufacturers, restrictions upon their trade, and taxation of their industry, were serious counts in the bill of indictment against the mother country. The blow they struck for equal rights . . . bequeathed us an enfranchised industry and respect for property, without which the useful arts can never flourish." (11, 9) Bishop recognized that the principle of economic freedom for the "useful arts" had been added to those of religious and political freedoms as foundation stones of the American republic.

The Americans drew unexpected support from the Englishman Adam Smith, whose monumental book *The Wealth of Nations* — published in 1776, the year of the Declaration of Independence — confirmed in theory what they had already put into action: that freedom of trade must replace defensive national mercantile systems. Smith warned his countrymen about the risk of attempting by law to "raise up a nation of customers who should be obliged to buy from the shops of our different producers, all the goods with which these could supply them." (81, 626) He proposed that enlightened self-interest had the power to reduce labor and trade to their purest relationship, whereby men may compete freely to provide products or services that others need. With all that this implies in terms of economy and quality, the principle of freedom of trade and its corollary, freedom of enterprise, defined the philosophy by which Americans generate products to meet consumers' needs and desires and then compete openly for their share of the market.

It became inevitable at this point that the practice of design as the organization of means toward predetermined ends would emerge as the essential link between the producers and consumers of the products of free enterprise. Moreover, the counterpoint of the related yet independent philosophies of the patricians and the Puritans created a cultural texture in which each was able to retain its sometimes complementing, sometimes contrasting character.

The Arts of Affluence

Daniel Henchman Takes this Method to inform his Customers in Town and Country, That he still continues to carry on the Gold and Silversmith's Business at his Shop opposite the Old Brick Meeting House in Cornhill, where he makes with his own Hands all kinds of large and small Plate Work, in the genteelest Taste and Newest Fashion . . . equal in goodness and cheaper than any they can import from London.

advertisement, *The Boston Evening Post*, 1773 (116)

In the new towns and young cities of America, a second stream of design consciousness emerged as the citizens sought to prove their cultural equality with the Old World. Wealth was understood in America as the only avenue by which one could achieve a degree of elegance equivalent to that ensured for the European aristocrats and nobles by inheritance and primogeniture. The rapid accumulation of wealth enabled the Americans to buy the best England had to offer or to commission work in the style of the moment by émigré and native artisans in America. Thus, the urban buildings and furnishings of the colonies took on the familiar or fashionable styles that were prevalent in England and Europe. The newcomer to a colonial town was confronted with a mixture of Dutch-style townhouses, English-style public buildings, and French furnishings that enhanced the quality of urban living in the New World, graced the social behavior of its citizens, memorialized their historic events, and flaunted a growing American affluence that presumed to be on a par with the best on the other side of the Atlantic.

The greatest distinction in the drive for cultural status among the newly rich of America went to those who were able to display the latest styles from abroad. No colonial gentleman and no artisan seeking patronage wanted to be left behind in the fashion race. Even those who declared publicly that America had no time for such frivolities steeped themselves privately in English and continental fashions. No less a patriot than Benjamin Franklin wrote to his wife from Europe advising her to follow the latest fashions in home furnishings. Later, while George Washington was in the field against the British, workmen were busily renovating his home at Mount Vernon using English style books for reference. There is an interesting paradox in the fact that the colonial style of furnishings that today is considered a near-sacred standard of permanent excellence was, in its own day, readily discarded as soon as a fresher style could be unloaded at the dock. The paradox is compounded all the more by the fact that, as much as Americans today may compete for the latest fallout of high technology, even more do they hold tight to their collective (or assumed) colonial heritage.

Some critics of American culture have maintained that, because the émigrés had renounced their national origins, they became culturally impotent and therefore unable "to produce a culture in which the arts could flourish." (77, vi) It has been suggested that the fine arts were reserved for the exclusive pleasure of the upper classes, who were obliged to display or distribute from time to time some small portion of their treasure as aesthetic alms to the lower classes. John Fiske proposed that Americans should accept the theory of the "transit of civilization"—that their culture had to come from abroad, and that they needed to develop "carriers" to bring the fine arts to America. Fiske contended that there was a "cultural lag" whereby an aesthetic fashion that emerged abroad would not become popular in America for some years after it had reached its zenith elsewhere, and that the length of the "lag" was determined by the clarity of the style, the cultural vigor of the movement, and the means of transport. During the colonial period, it was presumed that 20–30 years would lapse before a new European fashion would gain a foothold in America. Peaceful and prosperous times obviously quickened the flow, and, conversely, it seems that the greatest advances of original design in America have been made when the inflow of foreign cultural influence has been constrained by political and economic circumstances.

In the beginning the colonial Americans were content to reproduce the treasures that had been carried over from Europe. Not only implements but also pieces of furniture, pottery, glass, iron, copper, and brass products, pewter and silver wares, and fabrics of all kinds were used as patterns for duplication. However, as time went on, such products became dated in fashion and the wealthier patrons began to seek out and commission those craftsmen who could assure them that they were knowledgeable of the newer styles from abroad and competent in their manufacture. In this context, the advantage lay with those artisans who had emigrated the most recently, bringing with them samples and templates and patterns from which they could reproduce objects in the latest fashion. Journeymen artisans in the crowded shops abroad realized that their knowledge and experience would be welcomed in America and that they would be free, away from Old World guild practices, to set up their own businesses. Moreover, they were certain that young apprentices would be readily available to help them in return for being taught their craft.

For the most part the colonial craftsman did not consider himself to be a designer, but rather the instrument by which the desires of his patrons could be satisfied. To show his familiarity with the most recent styles, he imported examples that could be displayed to attract business and could also be copied. And he sought out and purchased special tools and patterns, or made his own from such samples as he could lay his hands on, that enabled him to work in the latest continental or English style. However, the artisan often found it necessary to modify a design—not only to suit a client's whim, but sometimes because of inadequate tools or limited talents. And the scarcity of labor in the colonies made it necessary to husband carefully the amount of energy that was put into a product by simplifying its form and ornamentation or by developing tools and methods that would demand less time. As a result, the products of the colonial craftsmen often achieved a taut refinement of form and a restraint of ornament that placed them above and beyond the extravagant originals. This attention of the colonial craftsman to labor-saving forms and procedures helped to refine the principle of economy of means as another of the basic principles of American industrial design.

It was expected in the colonies that a gentleman would be knowledgeable in culture and fashion as well as in science and philosophy. It was his obligation to see to the quality of his environment and to direct the character of the products made to his order by artisans. Educated men were presumed to know and understand architectural style and structure because they were often called upon to determine the forms of public buildings (as well as their own residences). Bruton Parish Church in Williamsburg, Virginia, was designed in 1711 by the governor of the colony, Alexander Spotswood, and Thomas Jefferson took a particular interest in the subject ("Architecture is my delight") and found time from a full career to design the main buildings of the University of Virginia as well as his own home, Monticello.

The basic design sources for Jefferson and the other gentleman architects of the colonies were the popular books of architectural drawings and illustrations of the time, which provided both gentlemen and builders with a ready reference to English and continental styles. Jefferson is known to have had at least five books on Andrea Palladio, who made extensive measurements of ancient Roman buildings and published their basic proportions as early as 1570. Palladio's flawless sense of proportion and sympathy for the classical style exerted a profound influence on Western architecture well into the nineteenth century, and Jefferson undoubtedly based Monticello on his work. The émigré Peter Harrison certainly followed the pattern book *Andrea Palladio's Architecture* (London: Edward Hoppus, 1735) in designing the facade of the handsome Redwood Library built in 1750 in Newport, Rhode Island. Harrison's main occupation at the time was in business in Newport and as a collector of customs at New Haven, Connecticut. However, he is sometimes identified as the first American architect because he was the first person on record known to have been paid a fee by a patron to design a structure to be erected by a builder. Harrison was paid 45 pounds for his plan for Christ Church Episcopalian (Cambridge, Massachusetts) in 1761. Before this he had designed

The Redwood Library in
Newport, Rhode Island,
planned by Peter Harrison
(perhaps with the assis-
tance of his brother Jo-
seph), was in all likelihood
based on a Roman Doric
temple design derived
from a book on Palladio.
Preservation Society of
Newport County.

Andrea Palladio: The Four
Books of Architecture,
Isaac Ware's 1783 repro-
duction of Palladio's orig-
inal plates and translation,
is generally considered
one of the best of the Pal-
ladian editions.

King's Chapel in Boston, for which he had been promised payment that was never made. It is believed that Harrison drew ideas for this church from the best-known architectural pattern book of the time, *A Book of Architecture,* by James Gibbs (second edition: London, 1739).

It is known that others, like Samuel McIntire, an architect-builder and carver of Salem, Massachusetts, depended upon Batty and Thomas Langley's *The City and Country Builders and Workman's Treasury of Design* (London, 1740). And Robert Smith, who had served as the master carpenter for Nassau Hall at Princeton University, designed and built Carpenter's Hall in Philadelphia, taking his concept from *Palladio Londinensis* (London: William Salman, 1734). Other pattern books on architectural style, such as those by William Halfpenny and Isaac Ware, undoubtedly found ready acceptance in the colonies. They not only provided the builder with aesthetic and structural guidance, but also stressed that a building should suit its function.

The pattern books had a strong effect on several aspects of design in the colonies. Any citizen could use them to take up a trade, or at least to duplicate a building or an object. Although they did not stimulate originality as much as imitation, considerable ingenuity was often required to modify the designs illustrated to accommodate the available talents, tools, and materials. Moreover, the general aesthetic consistency from pattern book to pattern book nurtured homogeneity in buildings, furniture, furnishings, and the other accessory arts. And the fact that the design drawings were reproduced in books of engravings and sold in quantity suggests that the origin of design as a profession in its own right, separate from that of the artisan, may have been stimulated by these pattern books. At the very least, they enhanced the artist-designer's reputation and earning power.

The earliest furniture and furnishings in America were undoubtedly those few treasured pieces that the émigrés had been able to stow aboard the small ships that brought them from the Old World. Although few authenticated pieces survive, there is an unbroken line of European and primarily English influence in American furniture. The Elder Brewster armchair, whose origins go back to Romanesque styles, serves as a good example of the "stick" type of chair that was used in the colonies in the mid-seventeenth century. This type of chair was constructed of turned oak and ash spindles jointed with the least possible effort; although ungainly in form and of questionable comfort, it was perfectly suited to the simple tools of the turners and joiners of its day. The earliest American tables and cabinets were influenced by the Flemish style of the sixteenth century as adapted to English tastes in Elizabethan and Jacobean England. Although their form was perhaps as heavy and clumsy as that of the original products, the decorative details were not cut so deeply into the wood as they were on the English and Dutch prototypes.

In the middle of the seventeenth century, after the restoration of the monarchy in England, a rich baroque style appeared as a reaction to the restraints of the Commonwealth. This "William and Mary" style, based on combined Italian and Dutch influences, was introduced to the colonies in 1700 when the new Royal Governor brought over pieces that exhibited cunning forms and richly patterned surfaces and showed the influence of Sir Christopher Wren.

John Gloag has aptly described the eighteenth century in Europe as the "magnificent century" of architecture and furniture design. "The character of magnificence," he writes, "varied with the country; in France, as society steadily advanced toward dissolution, it became frantic; in the etiquette-ridden states of Germany, oppressive; in Austria and Italy, gay; in England, restrained but consistently gracious." (38, 164)

Although Anne was queen of England for only twelve years (1702–1714), her name has been given to the era's most important style of furnishings. This more graceful fashion combined elements from France and the Netherlands into products in which form was considered to be more important than ornament. It was at this time that the cabriole leg was introduced from France. This graceful appendage, with its interlocked curves adapted from Greek and Roman sources,

In comparison with the unique hand-carved furniture of the aristocracy, this Brewster chair suggests by its lathe-turned parts and mechanical construction that it was a humbler manufactured and assembled product. It may even have been shipped across the Atlantic on the May-flower *in a knocked-down state.* Harper's Monthly, January 1877.

was fundamental to the Queen Anne style and predicted the rococo. It has persisted well into the twentieth century as a symbol of refinement, not only on furniture but also on manufactured appliances such as chafing dishes, stoves, and refrigerators. The Queen Anne style of furnishings did not appear in the colonies until after Anne's death; however, it persisted through the reigns of George I (1714–1727) and George II (1727–1750).

In 1700 about half of the furniture being sold in the colonies was domestic in origin, much of it manufactured by émigré jointers and cabinet-makers. However, by the time of the Revolution urban craftsmen had taken over the trade that had once offered the best market for English manufacturers. Craftsmen from Boston, Salem, Portsmouth, Newport, and New York all joined this lucrative craft, but it was the furnituremakers of Philadelphia who developed the highest-quality colonial American furniture in the expensive pieces they produced for their clientele of wealthy merchants and traders. The period between 1725 and 1750 may easily be considered the prime period of furniture design and manufacture in the colonies, as a unique style was beginning to emerge that combined the clear expression of Queen Anne with the comparatively spare elegance and restrained ornament of colonial craftsmen.

The European peasant tradition of "stick" chair-making found ready acceptance in America in two forms apart from the Brewster type. The ladderback chair, which appeared as early as the eleventh century in Europe, consisted essentially of a structure of straight or turned members fitted with a woven rush seat. The "Windsor chair," which probably originated in the latter half of the seventeenth century, employed turned spindles that were socketed into a solid wood seat shaped to fit the body. John Gloag claims that the Windsor chair, "by anticipating the technique of mass-production," is "the only article that has survived the industrial revolution unmarred." (38, 156) And, although Gloag regards the Windsor as "the national chair of England

as the rocker is the national chair of America,"
it should be pointed out that in England the Wind-
sors were known originally as "wheelright's"
furniture and considered to be suitable only for
provincial use, whereas in America they were
immediately popular when they first appeared in
1725 and came to be widely used in private
homes at every social level as well as in taverns,
inns, and public buildings. A set of Windsor
chairs used by the signers of the Declaration of
Independence is preserved in Independence Hall
in Philadelphia as a treasured national artifact.
The Windsor was durable, well suited in shape
and proportion to the human body, and eminently
preferred by Americans on the move since it
could be knocked down for shipping. Soon it was
being manufactured throughout the colonies,
and with increased competition the shape, proportions, and dimensions were refined until they
achieved that happy harmony of form and fitness
that makes for ultimate beauty in man-made
utilitarian products.

While the Queen Anne style was developing its
richer beauty in the colonies, the English, following the lead of what had been done in architecture, began to publish books of engraved
plates of furniture designs. The first, *A Universal
System of Household Furniture* (Ince and May-
hew, 1748), contained, according to the preface,
"above 300 designs in the most elegant taste,
both useful and ornamental." (209, 191) Although
there is no positive evidence that this book was
known and used in America, there is no doubt
that the series of furniture style books that followed were avidly studied there by urban craftsmen and their clients.

The most important English book on furniture
design was undoubtedly Thomas Chippendale's
Gentleman's and Cabinet Maker's Director
(1754). Chippendale's 200 plates of furniture
designs established a new style by combining
Chinese, French, neo-Gothic, and Palladian elements into an elaborate English rococo fashion
that swept away the more reserved Queen Anne
style. Despite criticism from some members
of his profession, Chippendale's style was
supported by *Hogarth's Analysis of Beauty,* which
praised its serpentine lines as the epitome of

The Windsor chair, exemplified by this high fan-back from Delaware, epitomizes the colonial style. Yet its technology is admirably suited to low-cost, high-volume production and economical storage and shipping. Index of American Design, National Gallery of Art.

The fanciful design variations for a chair shown in Thomas Chippendale's The Gentleman and Cabinet-Maker's Director *were a source of inspiration for furniture makers of lesser imagination in England and the colonies.*

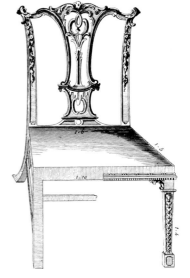

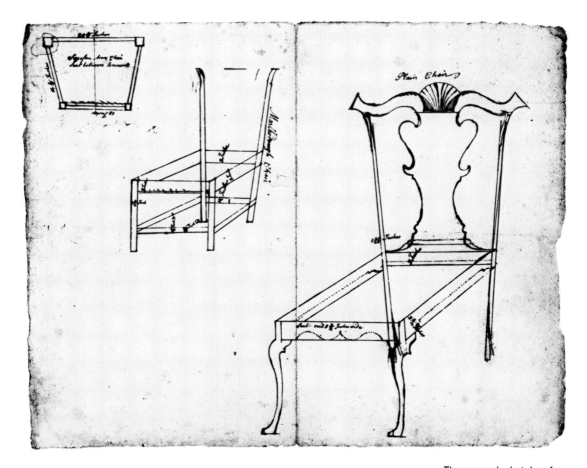

These rough sketches for
a side chair, from Samuel
Mickle's design book,
show Chippendale's in-
fluence in a transitional
concept combining Chi-
nese cresting with a solid
splat in the Queen Anne
style. Philadelphia Mu-
seum of Art; given by
Walter M. Jeffords. Photograph by A. J. Wyatt.

beauty. The English rococo style became as popular in the colonies as it was in England, and although no copies of the Chippendale book survive today in the collections of colonial books at least 29 copies are known to have been in the colonies prior to the Revolution, including one mentioned in the records of Carpenter's Hall in Philadelphia.

Thomas Chippendale's success as a conceiver and publisher of designs suggests that he, more than any other artisan-designer of his era, was a forerunner of the contemporary industrial designer. Although he was a competent craftsman, he elected to combine his talents with practical business methods and a sense of aesthetic direction in order to develop and market concepts that would meet the desires of his clients.

The style that Chippendale created was copied by others, such as Thomas Manwaring, whose *Cabinet and Chairmaker's Real Friend and Companion* (London, 1765) was also used as a source book by émigré and native cabinetmakers in America. Some, like Samuel Mickle of Philadelphia, kept notebooks in which they recorded their own adaptation of Chippendale's designs. The finest furniture in Philadelphia came in the decade just before the Revolution from the workshops of Randolf, Savery, Gostelove, Affleck, and other cabinetmakers who produced original concepts of furniture (such as the highboy) by recombining elements from Chippendale. However, by this time, feelings were running against English fashions and ideas, and American artisans took pains to convince their clients that their products were not imported but "made in America"—a fact that was just as good for business as it was for politics. Thus, the flow of English influence was set aside, at least publicly, until the end of the War of Independence, when the cultural influx from England to her late colonies began again as strongly as ever.

Colonial silversmiths (although they were called goldsmiths in their day, the title of silversmiths is more appropriate since they worked primarily in that white metal) had access to the various

This side chair, made in New York, follows the Chippendale style closely. However, it is clumsy in weight and proportion compared with its Philadelphia equivalents. Courtesy of Henry Francis du Pont Winterthur Museum.

architectural and furniture style books through their patrons. The primary design sources for their products, however, were actual pieces of continental and English silver, or at least patterns, drawings, and templates taken from the originals. For ornamentation, they either duplicated originals or borrowed freely from allegorical figures in prints or from cipher books known to be in the colonies at the time.

The silver of the colonies is interesting not only because of its high quality but also because of its important role in the economy of the colonies and its close association with outstanding events and personalities. There were goldsmiths in the company of the first ship sailing to Virginia in 1608, sent not to practice their craft so much as to search for precious metals in the New World. That they found none was confirmed by Captain John Smith. As other sources of wealth were developed, the practice of silversmithing was delayed until later in the seventeenth century. Until then, such silver as appeared in the colonies was brought from Europe in personal treasures (as has been confirmed by wills and inventories).

The first silversmiths known to have practiced in the colonies were Hull and Sanderson. John Hull (1624–1683) came to America when he was 11 years old after some training in the craft in England. When the General Court of Massachusetts established a mint in Boston in 1652, he was appointed master of the mint and selected another London-trained silversmith, Robert Sanderson (1608–1693), as his partner. Together they coined the first colonial shillings, originally with a willow tree imprint, then an oak, and finally a pine. Hull became immensely wealthy because he was permitted to keep one out of every twenty shillings for his service—undoubtedly the earliest and most remunerative form of a royalty contract in America. Hull and Sanderson also made the first known silver in the colonies by converting coin into plate. They affixed their mark to it for identification, rather than attempting to follow the more complicated system of English hallmarking. For over a century this practice of transforming coinage into silverware proved to be the most effective way of conserving wealth and

making it readily identifiable and redeemable. The colonists' mistrust of paper currency because it lacked fixed value created a demand for silver as a dependable form of currency. Silversmiths "established the mode of 'investing' their silver coin and that of their clients in the form of porringers, tankards and the like, for both immediate use and future security." (85, 4) Thus, because of their knowledge of the value of the coinage of the various nations and their ability to assay, melt, and fabricate the precious metals, they served (at least symbolically) as bankers, often achieving prestigious political and social positions as a result. Next to the clergy they were the most respected professionals in the colonies.

In general, silver plate was quicker than furniture to embrace changes in style because prototypes were more easily transported and colonial merchants were all too willing to accept as payment for their products silverware that carried the respected quality marks of the English Worshipful Company of Goldsmiths.

The earliest colonial silver reflected the baroque Flemish forms that were popular in Europe and England. However, the American copies were sturdier and more reserved than the originals and often achieved an elegance of proportion and a form and dignity appropriate to their function. The silver of New Amsterdam (as New York was known until 1674, when it was ceded to England by the Treaty of Westminster) showed a particular preference for the beaker form for both secular and ecclesiastical purposes. The form was widely copied in the colonies, and a number of examples have been preserved because they passed into church collections as communion vessels.

In 1685 Louis XIV revoked the Edict of Nantes, driving many of the French Protestants into exile, and when in 1687 he forbade the Huguenot artisans to practice their craft in France and ordered them to melt down their plate many emigrated to the colonies. Bartholomew Le Roux (1663–1713) in New York, Cezar Ghiselin (1670–1734) in Philadelphia, and Apollos Rivoire (1702–1754, the father of Paul Revere) in Boston were outstanding Huguenot silversmiths who rejected their French origins to work in the popular English styles of the colonies.

The lineage of American silversmiths, which began when John Hull (1624–1683) took on his first colonial apprentice, Jeremiah Dummer (1645–1718), continued without a break until the death of Paul Revere. Dummer's work bridges the early American silver styles from the Restoration through the William and Mary baroque into the Queen Anne period. His best work was in the rich play of gadrooned and fluted surfaces against the plain surfaces of the baroque that John Marshall Phillips believed ushered in "the golden age of American silversmithing at a time when the centers of the craft had yet to celebrate their centenary." (71, 61) Several of Dummer's surviving cups appear to have been made serially, with apprentices finishing them to suit the wishes of one patron or another.

One of Dummer's apprentices was his brother-in-law John Coney (1655–1722), generally considered the most sensitive of the colonial silversmiths. His masterpiece is undoubtedly the Monteith, which is now the prized centerpiece of the Garvan collection at Yale University. Its form, taken from an English prototype, illustrates the desire of the first generation of native-born New Englanders to offset their provincialism by acquiring fine, elaborate objects in the latest English style. In 1715, toward the end of his life Coney took on Apollos Rivoire as an apprentice, and Rivoire was still there in 1722 when Coney died. Rivoire subsequently trained his son Paul Revere (1735–1818) in the craft of silversmithing.

The William and Mary baroque style became so popular that many earlier pieces of silver plate were lost when they were melted down to be re-shaped into the new form. Eventually so much sterling coinage had disappeared into the melting pot to become silverware that the British Crown found it necessary in 1697 to raise the percentage of silver used in plate from the Sterling Standard of 0.925 used for coinage to the Brittania Standard of 0.96, thus increasing the cost of the basic material. The new alloy was, moreover, too soft for the thin elaborate forms of the

baroque, and so that style was slowly replaced by shapes that attempted to make up for their loss of ornament by refinement into more graceful lines and by the addition of stiffening bands and moldings. By 1720 the new style (to be known as Queen Anne) had established itself in the arts of luxury and the Brittania Standard was revoked.

The plainer Queen Anne forms of silver were most likely introduced to the colonies by the queen's donation of ecclesiastical plate to the major Episcopalian churches there. The style had a particular appeal to the colonists because of their suspicion of their association of the overly ornate with an oppressive aristocracy and their innate sympathy for the restraint of the Puritans. Even so, the plain surfaces invited the enrichment of fine engraving. In this period the fashion was to use elaborate ciphers of interlaced initials that were popular in England. Sympson's *New Book of Cyphers* (London, 1726) was known and used by colonial silversmiths such as Simeon Soumaine (1685–1750) of New York, and in John Singleton Copley's portrait of the Boston silversmith and engraver Nathaniel Hurd (1729–1777) one may see illustrated Guillim's *Display of Heraldry,* Hurd's source of armorial devices.

By 1730, while the Queen Anne style still dominated American silver, a new mode was spreading rapidly across all of the decorative arts in England. It was an exuberant rococo that had originated in France out of Louis XIV forms and ornament. Asymmetrical decorations embodying scrolls, plants, and shells in cast, chased, engraved, and pierced techniques all but obscured the original shape of the product.

In the midst of this period the great *Encyclopedie* of Denis Diderot appeared. The first volume was published in 1751 and the last in 1772. The sections on furniture and silver displayed the extravagant forms of the French rococo, and must have been a strong influence on the work of Thomas Chippendale, on that of Paul de Lamerie (an exiled Huguenot master silversmith working in London), and through Lamerie and others of his countrymen on the silver of England and the

The Anglo-Flemish style
is reflected in the form of
this sturdy seventeenth-
century beaker by Hull
and Sanderson. Its proto-
type may have been an
earlier Dutch beaker
known to have been in
Boston at the time. Yale
University Art Gallery,
Mabel Brady Garvan
Collection.

A standing cup by Jere-
miah Dummer in the
handsome William and
Mary baroque style. The
form was fashionable in
England at the time, and
it is likely that Dummer
made up the bowls first
and then fitted some with
stems (as illustrated) to be
used as communion chal-
ices and others with han-
dles to be used as secular
caudle cups. Courtesy of
Museum of Fine Arts,
Boston.

The Monteith, by John
Coney, is a monumental
example of American sil-
ver in the William and
Mary baroque style. It is
better proportioned than,
and as lavish as, its En-
glish prototypes. The scal-
loped rim was used to
carry glasses, with their
bowls immersed in cold
water, into the room where
the ceremonial punch was
to be made. Yale Univer-
sity Art Gallery, Mabel
Brady Garvan Collection.

*Drawing of a Monteith that
was owned by the Vint-
ners' Hall in London in
1702. Joseph Wilfred
Cripps,* Old English Plate.

Jacob Hurd's masterpiece was this two-handled covered cup, presented in 1744 by several grateful Boston merchants to the commander of a colonial warship in an early naval victory over the French in King George's war. Its simpler Queen Anne style reflects the British Crown's increase in the percentage of silver in coinage. Yale University Art Gallery, Mabel Brady Garvan Collection.

A page of designs for silver forms in the rococo style conceived by artists and engravers for the Diderot Pictorial Encyclopedia of Trades and Industry (volume II). They were modified and reproduced, and they served as a source of ideas for expatriate Huguenot silversmiths and other craftsmen in other lands, including the colonies.

colonies. Although the *Encyclopedie* illustrated the luxury of the rococo, it was in itself the first great effort to acknowledge and to gather into a single source all of the knowledge that was coming from the first stirrings on the continent of intellectual tolerance and rationale. By exalting the acquisition of scientific knowledge, Diderot, with the help of his collaborator D'Alembert, spread the democratic concept that government existed at the pleasure of and for the good of the people. In a way, the *Encyclopedie* presaged and may have accelerated the revolutions that were to come, first in the colonies and later in France. Most important, it signaled the dawning of the age of technology.

For a while, the common effort of the English and their colonials in the French and Indian Wars (1756–1763) drew the two closer together, and the lively trade that ensued encouraged the importation of much English silver in the new rococo style. Some colonial silversmiths became importers of plate; others resented the threat to their own livelihood.

The surviving letters, notebooks, and journals of colonial artisans occasionally include illustrations of products or product details, but these appear to be records of work done rather than original designs to be followed. One design book that escaped the attrition of time is that of the Annapolis silversmith William Faris (1728–1804), which is now in the collection of the Maryland Historical Society in Baltimore. It contains drawings of various silver objects in the rococo style that were apparently traced from templates and patterns made either from the original pieces or purchased from another source. The records of Paul Revere, Jr. (1735–1818), now in the collection of the Massachusetts Historical Society in Boston, show tracings that were probably made from the original pieces as well as other rococo decorative patterns. Another useful source of information about the practice of colonial craftsmen is the written inventories that have been preserved. It was customary in early America when an artisan died for the court to appoint a committee including members of the deceased's trade to make a detailed inventory with a realistic appraisal of his tools, materials, and uncompleted projects.

The most elaborate sample of the rococo style in the colonies is a kettle and stand made in Philadelphia by Joseph Richardson, Sr. (1711–1784). Despite its extravagance of form and ornament, however, it was only a modest interpretation of the unrestrained English and French work of the period. Another example of Philadelphia rococo of unusual historic interest is the inkstand or standish, fashioned by Philip Syng, Jr. (1703–1789), that was used by the signers of the Declaration of Independence in 1776 and again in 1787 by representatives of the colonies adopting the Constitution. It is now in Independence Hall in Philadelphia.

In New England, the rococo influence manifested itself primarily in an elaboration of decorative detail rather than an exaggeration of form. It was also not uncommon for silversmiths to transpose the forms of Chinese porcelain (or English copies thereof) into silver objects. The "Sons of Liberty" bowl made by Paul Revere in 1768 is said to have followed a ceramic prototype. There is a porcelain bowl of this type in the British Museum upon which is painted a portrait of John Wilkes, the English member of Parliament who supported the right to self-government in the colonies. Certainly Paul Revere used the oriental ceramic pitcher form as reproduced in England as a source for his own silver pitchers.

Even before the American Revolution, a rediscovered classical style was sweeping Europe that was to furnish a stylistic base for the buildings and furnishings of the new republic. Once again American fashion was to fall into line behind European sources and to begin with them a backward spin through the Greek agora and the Roman forum, into Egypt and the bazaars of the Persians, seeking cultural justification.

Daniel Henchman

Takes this Method to inform his Customers in Town & Country, That he still continues to carry on the GOLD and SILVERSMITHS Business at his Shop opposite the Old Brick Meeting House in Cornhill, where he makes with his own Hands all Kinds of large and small Plate Work, in the genteelest Taste and newest Fashion, and of the purest Silver ; and as his Work has hitherto met with the Approbation of the most Curious, he flatters himself that he shall have the Preference by those who are Judges of Work, to those Strangers among us who import and sell English Plate, to the great Hurt and Prejudice of the Townsmen who have been bred to the Business.—— Said HENCHMAN therefore will engage to those Gentlemen and Ladies who shall please to employ him, that he will make any Kind of Plate they may want equal in goodness and cheaper than any they can import from London, with the greatest Dispatch.

Colonial newspapers carried advertisements similar to this one (from the Boston Evening Post, 1733) in which Daniel Henchman promises to make products at least equal in quality to those being imported. Courtesy of Massachusetts Historical Society.

*Design for a coffeepot
in the rococo style from
the design book of William
Faris. The main forms
were apparently traced
from templates, with
elaborate details added
in a fanciful manner.
Maryland Historical
Society.*

*Tracings of large and small
spoons made by Paul Re-
vere, Jr., for record purposes. Such tracings may
have also served as pat-
terns for other pieces.
Courtesy of Massachu-
setts Historical Society.*

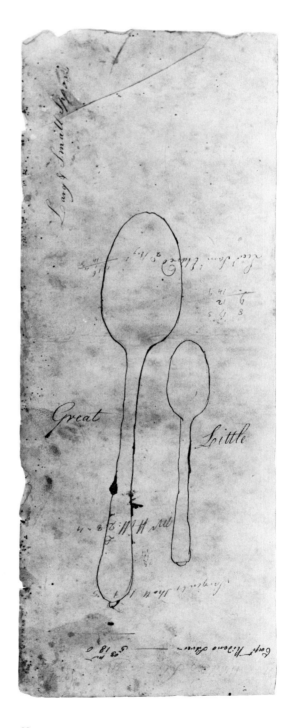

The first and last pages of the inventory of the tools, materials, and unfinished work of Benjamin Hurd (1739–1781). This record provides useful evidence of the methodology Hurd followed in his craft. Reproduced with the permission of James M. Connolly, Register of Probate, Suffolk County, Massachusetts.

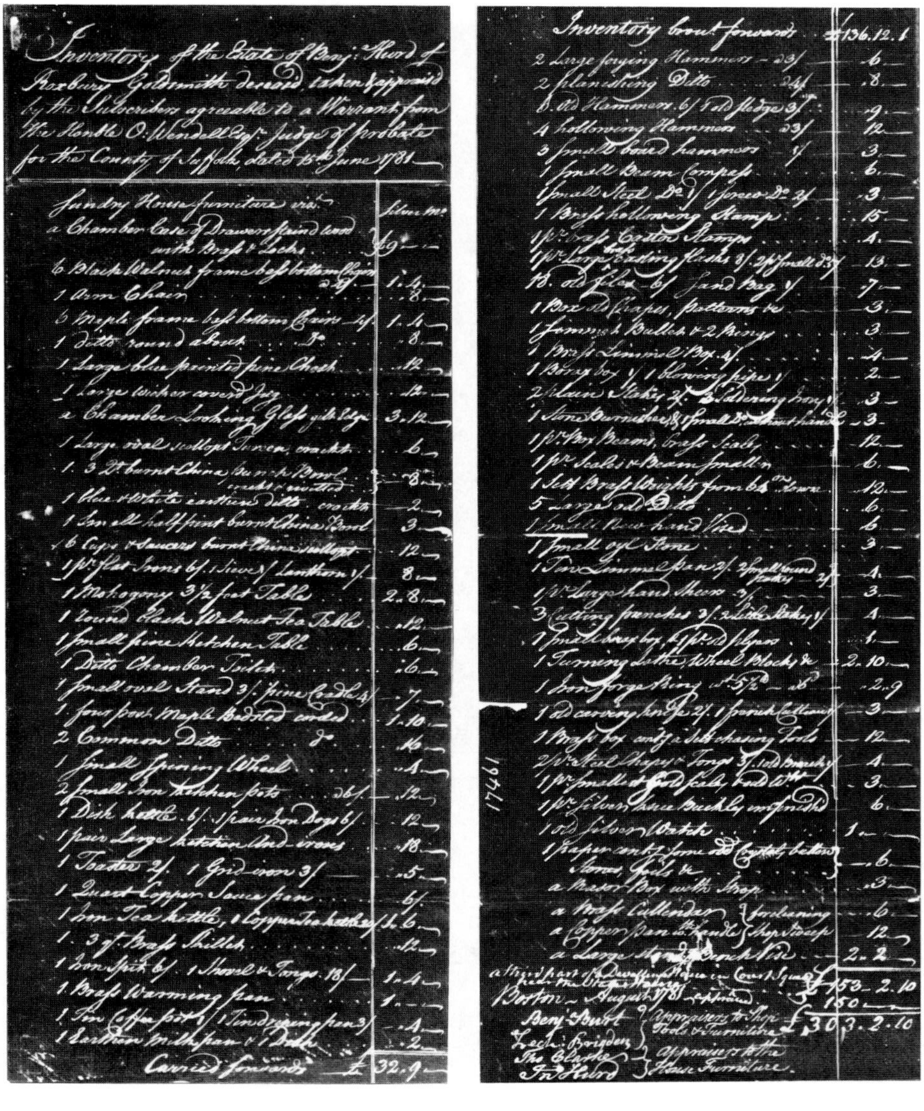

Inventory of the Estate of Benj: Hurd of Roxbury Goldsmith deceased taken & appraised by the Subscribers agreeable to a Warrant from the Honble O: Wendell Esqr Judge of probate for the County of Suffolk, dated 15th June 1781

Inventory brot forward £136.12.1

The only rococo teakettle and stand surviving from the colonial period, made by the Quaker silversmith Joseph Richardson, Sr., of Philadelphia. Although unusually elaborate for a colonial product, it is more restrained than its proto-type, made by Paul de Lamerie in England. Yale University Art Gallery, Mabel Brady Garvan Collection.

A creamware ceramic pitcher manufactured in Liverpool around 1800, primarily for export to the United States. Its form was based on an oriental prototype, and its ornament (the recent Revolution notwithstanding) was intended to appeal to the new nation. Yale University Art Gallery, Mabel Brady Garvan Collection.

A silver pitcher by Paul Revere based on the Liverpool pitcher. It was eminently suited to production after rolled sheet silver had become available. Yale University Art Gallery, Mabel Brady Garvan Collection.

This Federal sterling silver teapot, covered sugar bowl, and cream pitcher by Daniel Van Voorhis, with a tea caddy made to match by W.B.&T., show characteristic oval, urn, and helmet shapes. The teapot and the spout are fabricated from flat-rolled silver sheet. The ornamentation includes engraved swags of tasseled drapery in the Adams fashion. Courtesy of Art Institute of Chicago.

The Search for Identity

The expansive future is our arena. We are entering on its untrodden space with the truth of God in our minds, beneficent objects in our hearts, and with a clear conscience unsullied by the past. We are the nation of human progress and who will, what can, set limits on our onward march? . . . The far-reaching, the boundless future, will be the era of American greatness. . . .

New York State Representative John Louis O'Sullivan, 1780s (123, 8)

As one of the first groups of Europeans to seek to establish an independent nation, the Americans were anxious to confirm their identity. Now that they had cast themselves adrift from their heritage, some warned that this unprecedented social odyssey would not be able to survive without an antiquity to be revered, defended, and transmitted. Some expressed a fear that the new republic may have been born impotent and that only by some form of divine intervention would it be able to achieve a sustaining vitality. Others found comfort in a conviction that a uniquely American philosophy was emerging that promised that the past was irrelevant to the new nation. They argued that the Americans did not need ancestors because they themselves were ancestors, thus echoing John Locke's statement that "in the beginning all the world was America." (57, 319) The citizens of the young republic thus declared themselves independent of national origins and obliged to place their faith in individual conscience set against new challenges. In what was to become a strong New England movement a few years later, the American Transcendentalists counted on the presence of the Divine in each person as a source of truth and a guide to action.

The new citizens therefore sanctified every incident as a substitute for European antiquity. The landing of the Pilgrims at Plymouth Rock, the first Thanksgiving, the Boston Tea Party, and other events were hallowed as sacred moments of the new nation. Shrines were erected at the bridge where the embattled farmers fired "the shot heard 'round the world," on the hill where the barricaded patriots were ordered not to fire "until you see the whites of their eyes," and in the woods where George Washington prayed for divine guidance.

In compensation for the loss of their European roots, the disenfranchised colonists hastened to embrace every symbol that would help them create an instant history of their own. From Benjamin Franklin's political cartoons to the ancient political and architectural forms that were adopted, every distinctly American aesthetic abstraction was cherished. These symbols still permeate the communal and decorative arts of the United States.

Every product that had figured in the struggle for independence acquired an historic patina and was destined to become a model for countless reproductions in the future. The silver bowl that Paul Revere made on the eve of the Revolution for the Sons of Liberty in honor of the 92 patriots who voted not to rescind their letter urging the other colonies to unite against the British is the most honored relic of the time. The Windsor chairs and the silver inkstand that were used in the signing of the Declaration of Independence are displayed proudly in Independence Hall. Even Franklin's glasses, Washington's surveying instruments, and Jefferson's drawing tools have become venerated relics of the American political saints.

In what must have been the first corporate identity program ever undertaken in the country, the leaders of the young republic ordered the design and development of heraldry and instruments and monuments of state. They resolved that a national banner be sewn, that a seal be devised, that coinage be struck in the Roman decimal system, and that the fashionable orders of ancient Greece be adopted for official architecture. For designers these acts mark the origin of style employed with deliberate intent to define and project the philosophy and ideals of a client. Two of the symbols, the flag and the Great Seal, clearly illustrate the two ends of the spectrum of corporate identity—the flag as a natural product of inevitable evolution and the seal as the result of deliberate invention.

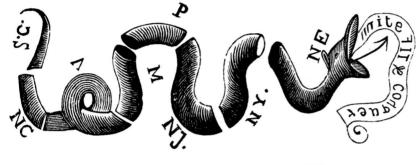

JOIN or DIE

This popular newspaper heading, devised by Benjamin Franklin in 1754 to urge the English colonies to unite against the French, was revised in 1776—this time against the British. Harper's Monthly, December 1875.

A second editorial cartoon by Franklin, drawn in 1774 and depicting, in his words, "the Colonies (that is, Brittania's limbs) being severed from her. . . ." Harper's Monthly, *July* 1875.

On the bowl:
To the Memory of the glorious NINETY-TWO: Members of the Hon.ble House of Representatives of the Massachusetts-Bay, who, undaunted by the insolent Menaces of Villains in Power, from a strict Regard to Conscience, and the LIBERTIES of their Constituents, on the 30th of June 1768, Voted NOT TO RESCIND.

The Sons of Liberty Bowl (of which countless reproductions have been marketed as patriotic tributes to the Revolution), fashioned by Paul Revere, Jr., in 1768. Courtesy of Museum of Fine Arts, Boston. Gift by subscription and Francis Bartlett Fund.

Despite its romanticized history, the American flag followed a clear path of evolution from (1) the English sign of St. George and (2) Scotland's emblem of St. Andrew to (3) their combination in the Union Jack, or King's Colors. Then (4) their cojoining in a union with corner of a crimson banner as the Union Flag was followed by (5) the breakup of the crimson field into red and white stripes for the thirteen colonies in the Grand Union flag, and, finally, (6) the replacement of the two crosses in the field with thirteen stars to represent each of the United States. Mark Waterman.

The first legislative action on record establishing the national flag was taken in June 1777, when the Continental Congress "resolved that the flag of the thirteen United States be thirteen stripes alternate red and white, that the union be thirteen stars, white on a blue field, representing a new constellation." Despite the popular stories of the roles of George Washington and Betsy Ross in the development of the flag, the facts are that its general design was a natural extension of preceding English flags. Only the origin of the decision to substitute stars for the combined crosses of Saint George and Saint Andrew seems to be lost in history. The ancient sign of Saint George (a red cross on a white field) had served as the banner of the English from the fourteenth century until 1606, when King James I, by royal proclamation, ordered that it be united with Scotland's sign of Saint Andrew (a white diagonal cross on a blue field) as evidence of an alliance of the two kingdoms. The King's Colors, as it was called, was ordered to be displayed from the maintops of the king's own ships. (Since other English ships were still permitted to carry Saint George's red cross, it is quite likely that the Mayflower displayed that banner when she landed the Pilgrims at Plymouth.) In 1707, after ratifying the complete union of England and Scotland, Parliament ordered that the two crosses should be cojoined in the upper inner corner (the union) of a crimson banner. This new flag was commonly known as the Union flag.

The Union flag was used in America with other local banners until the colonies separated from the mother country. When George Washington arrived in Cambridge, Massachusetts, in July 1775 after the battle of Bunker Hill, this Union flag was flown over his camp, still acknowledging (at least as far as the British believed) allegiance to England. It was during this encampment that the Continental Congress decided that the troops raised by the several colonies to oppose the English should henceforth serve as the combined forces of "the United Colonies of North America."

In October 1775, when the Continental Congress authorized the establishment of the first Federal Navy, there was still no national flag for the 17 ships to sail under. However, by the end of the year a committee including Benjamin Franklin, Thomas Lynch, and Benjamin Harrison recommended that the crimson field of the English Union flag be changed to "thirteen stripes, alternate red and white, emblematic of the union of the thirteen colonies." (72, 218) This Grand Union flag, still bearing the English and Scottish crosses in the union, was raised for the first time in January 1776 over Washington's camp at Cambridge. The British, and perhaps even most of the Americans, did not realize yet that the new flag was symbolic of the struggle for complete independence that was to come. By the middle of the year it was flown (in Washington's presence) at the reading of the Declaration of Independence in New York. After that, Washington had it flown over his fortifications and headquarters.

The Grand Union was the national flag of the Americans in September 1776, when Congress ordered the words "United States" be employed "where heretofore the words 'United Colonies' had been used." Despite the Congressional resolution of 1777 establishing the stars and stripes, there is no evidence that General Washington or the armies of the United States ever used the new flag until 1783, after the Revolutionary War was over. The first flag of the United States with 13 stripes and 13 stars was flown until 1795, when they both were increased to 15 in recognition of the addition of the new states of Vermont and Kentucky to the Union. In 1815, as other states joined the Union, a commission recommended that the stripes be reduced to the original 13 and that a new star be added in the future in honor of each new state that joined the Union. In 1818 this resolution was passed by Congress.

For national colors the Americans had selected the red, white, and blue of their mother country, England, and of France, their great ally in the War of Independence. Somewhat after the fact, the first issue of *Columbian Magazine* in 1786 ascribed poetic value to the colors: "White signifies purity and innocence; red, hardiness and valour; and blue, the color of the chief, signifies vigilance, perseverence and justice." (By some form of subconscious allegiance, the citizens of many countries show a preference for their national colors in applications that are not necessarily patriotic. More often than not the colors red and blue, with white as the background, are selected by Americans for the identity programs of major transportation, utility, and energy companies, and blue alone—red having been preempted by another political philosophy—appears with unsurprising frequency as the corporate color of major American manufacturers.)

On the same day that the Declaration of Independence was read, July 4, 1776, Congress appointed Benjamin Franklin, John Adams, and Thomas Jefferson to devise the official seal that would bind all of the colonies (soon to become states) to commitments made in the name of the Union. The good intentions of this committee were dissipated in an effort to agree upon the appropriate symbolism for the new nation. Franklin insisted upon a historic analogy: Moses parting the Red Sea as the Pharoah and his legions were overwhelmed by the waters. John Adams argued that an illustration of Hercules resting on his club after his great labors would be more logical. And Jefferson proposed that the seal should depict the citizens of the united colonies as the children of Israel in the wilderness, guided by a glowing cloud by day and a pillar of fire at night. With the assistance of du Simitiere, a French-Indian silhouette cutter and painter of miniatures, Jefferson combined all of the ideas into a proposal that was laid before Congress in August 1776. Congress did not accept this complex attempt to satisfy everyone. It did, however, introduce two elements, the eye of providence in a radiant triangle and the motto *E pluribus unum,* that were to appear on the final design.

After the rejection of the first seal design, two more unsuccessful attempts under other committees were made to satisfy Congress. Then in 1782, after the third proposal for the great seal had been rejected by the Congress, the whole matter was referred to its secretary, Charles Thomson, who called on William Barton to recommend a design. Although Barton's first proposal was elaborate and impracticable, Thomson was able to draw from it key elements for the final design. He accepted Barton's proposal for the reverse side as "a pyramid unfinished—in the zenith, an eye in a triangle, surrounded with a glory"—a motif signifying strength and duration. (122, 33) The main face elements of Barton's complex design were discarded except for the eagle that had served as a finial for the shield.

Barton's final design, with Thomson's revisions, was accepted by Congress in June 1782 as "the Device of the Armorial Achievement appertaining to the United States," in which "the escutcheon or shield is borne on the breast of an American eagle, without any other supporters, to denote that the united states of America ought to rely on their own virtue. . . . the American eagle is displayed, holding in his dexter talon an olive branch, and in his sinister a bundle of thirteen arrows . . . and in his beak, a scroll inscribed with this motto—E pluribus unum . . ." and "the olive branch and arrows denote the power of peace and war, which is exclusively vested in Congress." (122, 33)

The number thirteen (the original number of colonies) dominates the seal. There are thirteen arrows. There are thirteen leaves and thirteen berries on the olive branch. There are thirteen bars on the escutcheon and thirteen stars in the glory over the eagle's head. On the reverse, the unfinished pyramid has thirteen layers of stone and the motto Annuit coeptis ("God has favored our undertaking") has thirteen letters. The Great Seal of the United States has gone through several modifications since its adoption, the most recent one done by the Tiffany Studios in New York City in the late 1800s.

Although the design of the Great Seal had transformed the aggressive imperial Roman eagle into one symbolizing the protection deemed essential to the young republic, not everyone was satisfied. Benjamin Franklin complained in a letter to his daughter, with whimsical petulance, about the selection of the eagle—he preferred the turkey. "For my own part," Franklin wrote, "I wish the bald eagle had not been chosen as the representative of our country; he is a bird of bad moral character; he does not get his living honestly. . . . Besides, he is a rank coward. . . . He is therefore by no means a proper emblem for the brave. . . ." (36, 134)

To Franklin, as to other leaders of the young nation, it seemed necessary to begin the voyage into the political unknown by discarding whatever affection they may have had for England. He observed that "all things have their season, and with young countries as with young men, you must curb their fancy to strengthen their judgment." (77, 3) Yet after the War of Independence, Benjamin Franklin moved comfortably in the high society of England and France even while such homilies were encouraging the simple life at home.

Thomas Jefferson expressed a similar concern about the influence of the Old World on young Americans. A youth should not be sent abroad, he wrote, because "he acquires a fondness for European luxury and dissipation and a contempt for the simplicity of his own country; he is fascinated with privileges of the European aristocrats, and sees with abhorrence the lovely equality which the poor enjoys with the rich in his own country: he contracts a partiality for aristocracy or monarchy; he forms foreign friendships which will never be useful to him, and loses the season of life for forming in his own country those friendships which of all others are the most faithful and permanent. . . ." (51, 636) Yet Jefferson studied architecture abroad after the war and sent home architectural models and books to guide his countrymen.

The temper of the post-Revolutionary period called for the leaders of the young republic to take a pious attitude toward the European image of luxury, aristocracy, elegance, and sophistication. George Washington, despite his personal affection for comfort and luxury, took a public position as champion of the simpler essentials of life. In order to emphasize the need for national austerity, this southern gentleman, whose garments had previously been tailored abroad, elected to wear a suit of Connecticut broadcloth for his second annual message to Congress in 1790. In this respect he was echoing the earlier position of Samuel Adams, who never wore or permitted his family to wear English clothing. Adams enthusiastically advocated boycotting English products as an effective way of protesting

William Barton's first design for the Great Seal was overladen with heavy visual and heraldic symbolism. However, it displayed the escutcheon and eagle and the unfinished pyramid and eye that would be selected for the final seal. Reference 72.

The front of the seal drawn by William Barton at Charles Thomson's suggestion, accepted with minor changes by the Congress. Although simpler than some thought it should be, it contains all of the essential symbolism. Cigrand, Story of the Great Seal of the United States.

The reverse side of Barton's seal. It remained essentially the same except that the motto Deo Favente was replaced by Annuit Coeptis and Novus Ordo Seclorum was added around the base of the pyramid. Cigrand, Story of the Great Seal of the United States.

English restrictions on American manufactures. The second president, John Adams, was even more emphatic in his denunciation of the arts as the product of decadent societies and therefore dangerous fruit for a young country. He argued that in a democracy the people would be too busy earning a living to indulge in artistic affairs and, furthermore, that the Americans would not as yet have attained a level of sensitivity that would enable them to produce works of any significance.

This avowed suspicion of the arts as antithetical to the primary needs of the country made it psychologically easy for the national leaders to turn the attention of citizens away from the actual and fancied luxuries of their former rulers. However, although John Adams's puritanical zeal may have been original with him and his contemporaries, it is much more likely that they drew philosophical support from rebel European intellectuals such as Rousseau, who wrote in *Emile* that luxury and bad taste were inseparable, and that styles were set by the rich in order to show off their wealth and by the artists in order to take advantage of it. John Adams's position was even stronger. He questioned whether it was even "possible to enlist the fine arts on the side of truth," because in his opinion they had been prostituted by superstition and despotism.

Presumably, Adams was convinced that this was a time not for self-indulgence, but for general austerity acknowledging the state of poverty that followed the Revolution. "It is not, indeed," he wrote his wife Abigail from Paris, "the fine arts which our country requires: the useful, the mechanic arts, are those which we have occasion for in a young country as yet simple and not far advanced in luxury, although perhaps much too far for her age and character. . . . I must study politics and war, that my sons may have liberty to study mathematics and philosophy . . . geography, national history, and naval architecture, navigation, commerce and agriculture, in order to give their children a right to study painting, poetry, music, architecture, statuary, tapestry and porcelain." (1, 381)

Yet, in what Constance Rourke called the "fable of contrasts," however much America's leaders disavowed in public the cultural diversions that preoccupied the patricians of the Old World, they realized in private that, until a native American culture emerged, Europe would remain the fountainhead of aesthetic expression. They understood that even a young democracy would have need of monumental architecture, ceremonial furnishings, and heroic sculpture and paintings. For them and others of their day, not unlike those in contemporary social democracies, the fine arts were valued less for artistic expression than for historic content. The portraits of John Singleton Copley (1738–1815), Gilbert Stuart (1755–1818), and Charles Wilson Peale (1741–1827) reflect the intense idealism of the times. The panoramas of John Trumbull (1756–1843) and Benjamin West (1738–1820) glorified historic events of the colonists and their revolution. The next generation of artists would be caught up by an emerging technology and would combine invention and expression as their contribution to American culture, but for the moment patriotic symbolism prevailed.

Thomas Jefferson provides an excellent example of the conflict between practical necessities and aesthetic desires that prevailed in the minds of America's leaders during the country's infancy. Although he would reminisce later in 1825 that the first object of young societies was food and clothing, in his *Notes on the State of Virginia* (written between 1781 and 1783) he lamented that the first principles of the art of building were unknown in America. Furthermore, he wrote to James Madison that he was not ashamed of his enthusiasm for the arts, as their object was "to improve the taste of my countrymen, to increase their reputation, to reconcile to them the respect of the world, and to procure them its praise." (32, I, 433) It was while serving as Franklin's successor as minister to France from 1784 to 1789 that Jefferson expanded his interest in democracy to include science and technology, the study of the fine arts and architecture in particular. Upon his return to the United States, Jefferson designed and had built Monticello, and equipped it with many of his own inventions. Later, as mentioned above, he was to design

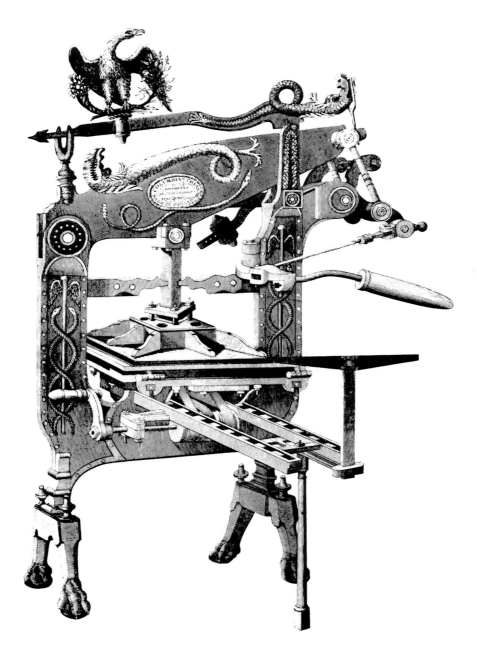

The Columbian Press, invented by George Clymer of Philadelphia in the early years of the republic, exemplifies the zealous use of symbolism by Americans anxious to establish a national identity. Abraham Rees (ed.), The Cyclopedia: or Universal Dictionary of Arts, Sciences and Literature.

the main buildings of the University of Virginia. As the most prominent patron of the useful arts of the period, Jefferson was to play an important part in directing and shaping the American identity.

It was inevitable that the Americans should reject the elaborate architectural style associated with Georgian England and embrace the spirit of classical form that was sweeping Europe. They sensed in its serene orders a provident complement to the cultural needs of the American republic. The émigré English architect Benjamin Henry Latrobe (1764–1820) applauded the choice of the forms of ancient Greece and Rome as the Federal style of the United States because, as he put it, "The history of Greece refutes the vulgar opinion that the arts are incompatible with liberty. . . . Greece . . . lost her freedom only when she prostituted the fine arts to the gratification of vice." (58, II, 205)

The most important influence in promoting the Federal style was in all likelihood that of Pierre Charles L'Enfant (1754–1825), the son of a Gobelin tapestry weaver, who had come to America in 1777 when he was 23 years old to serve as an engineer in the Revolution under Lafayette. His first design assignments after the war included the eagle emblem of the Society of Cincinnati, an altar screen for St. Paul's Chapel in New York, and a gigantic banquet pavilion for the New York legislature's celebration of the adoption of the federal Constitution. In 1787 L'Enfant remodeled New York's old city hall, originally built in 1699, to serve the national government. (George Washington was inaugurated as president on this building's balcony, and L'Enfant's design would serve as a model for the reviewing stand at Franklin Delano Roosevelt's inauguration in 1933.) The Federal style of government buildings was to persist for a century and a half.

When George Washington established the location for the District of Columbia in 1791, he selected L'Enfant to develop a plan for the new capital city. L'Enfant's design, in which the basic plan of Versailles was superimposed with a grid pattern, was an astute combination of imperial and democratic schemes. However, he pursued his assignment with such zeal that he alienated the commissioners who had been selected to supervise the project. As a result, just before the cornerstone of the new Capitol building was to be laid in a Potomac pasture, Washington was obliged to dismiss him. Later L'Enfant worked on the plans for Paterson, New Jersey, which Hamilton and other promoters of the Society for Useful Manufactures were founding as a factory town.

Charles Bulfinch (1763–1844) met Jefferson in Paris when Bulfinch was studying architecture, and returned to America to become a leading exponent with Jefferson of the Federal style. Bulfinch designed more than 40 churches and public buildings in New England, including the Massachusetts State House and Harvard's University Hall. Still, he considered himself a gentleman first and an architect second; to him, architecture was but one of the accomplishments of a gentleman. Bulfinch failed financially as an architect because in "reaching desperately for perfection in the art of architecture, he quite forgot that America at the dawn of the nineteenth century was a mechanic's rather than an artist's paradise." (3, 99)

When, in 1792, the Commissioners of Federal Buildings announced a public competition for designs for the national Capitol and the president's house, they were acknowledging the contention that existed among the gentlemen who considered that their classical education in history and style qualified them as architects, the émigré architects who had received professional training in their homelands, and the carpenter-builders who were experienced in the various arts of materials and construction. The idea for an open contest had been originally recommended to Washington by Jefferson in order to encourage the development of an architectural character that would be uniquely appropriate for the new nation. However, since the contestants had access to the same builders' handbooks and architectural style books, it was natural that the designs submitted would be classical in character, with echoes of earlier colonial buildings and decoration often inspired by American motifs.

FEDERAL HALL
The Seat of Congress
Printed & Sold by A. Doolittle New-Haven 1790

Peter Lacour delin. *A. Doolittle Sculp.*

The form of Federal Hall in New York City, the first capitol and the site of George Washington's inauguration in 1789, illustrates the classical influence of Palladio overlaid with L'Enfant's remodeling enrichments. New York Public Library, Stokes Collection.

The design competition for the Capitol was won by the gentleman architect Dr. William Thornton (1759–1828). Five years earlier Thornton, who was born in the British West Indies and educated in England, had won the first architectural competition held in the United States with his design for the Philadelphia Public Library. However, between 1793 and 1828, when the Capitol was completed, the architects Stephen Hallet and George Hadfield, the architect-engineer Benjamin Latrobe, and the gentleman architect Charles Bulfinch all took turns in supervising its construction and the many modifications that were made. The Irish-born-and-educated architect James Hoban (1762–1811) won first prize for a design for the president's house (later to be called the White House) in a competition that included, among others, an entry submitted by Thomas Jefferson under a pseudonym.

Benjamin Latrobe, of all of the designers practicing in the early days of the republic, comes closest to the contemporary ideal of a comprehensive designer. Latrobe's sensitivity to the line between the gentleman architect and the professional architect is evidenced in his letter to Henry Ormond in 1808: "I believe I am the first who, in our own country, has endeavored and partly succeeded, to place the profession of architect and civil engineer on that footing of responsibility which it occupies in Europe. But I have not so far succeeded as to make it an eligible profession for one who has the education and the feelings of a gentleman. . . ." (58, II, 67)

Latrobe was not only an engineer and architect, educated at the University of Leipzig with experience in practice in England, but also an accomplished watercolorist and furniture designer with a strong sense of ornament and style. He came to Virginia in 1796 to become engineer of the James River and Appomattox Canal. In 1800 he designed the Schuylkill River Waterworks to supply the city of Philadelphia, the first such water system in the United States. The system that he proposed depended upon the use of two low pressure rotative steam engines built in America by Nicholas Roosevelt after the plan of Boulton and Watt. Latrobe's interest in the potential of steam power, which was then dra-

matically new, led him to design a steam engine for the Washington Navy Yard and to attempt (unsuccessfully) to build a steamboat in collaboration with Robert Fulton.

In addition to his engineering accomplishments, Latrobe was quite aware of the waves of architectural fashion from England that were washing the shores of the new republic. His architectural achievements range from the Roman Catholic Cathedral in Baltimore (the first cathedral based on the Roman plan to be erected in the United States) to his Parthenon-based losing designs for the Second Bank of Pennsylvania in Philadelphia to a mansion for William Crammond in the Gothic style. (Wayne Andrews credits the popularity of the Greek style to the publication in England in 1762 of the first volume of Stuart and Revelt's *Antiquities of Athens,* and suggests that the interest in the Gothic style was stimulated by the Gothic castle that Horace Walpole built near London in 1750.)

After 1800, Thomas Jefferson, as the third president of the United States, appointed Latrobe to survey the public buildings in Washington. He made alterations to the White House, remodeled the Patent Office, and designed the south wing of the Capitol. After the Capitol was burned by the British in the War of 1812, Latrobe rebuilt parts of it, including the Hall of Representatives and the Senate Chamber. In the process he designed three column capitals in the Corinthian style, using as motifs the indigenous plants of corn, tobacco, and cotton. He also designed and had built a set of chairs and other pieces of furniture in the Greek style for the Blue Room of the White House. They were most likely built by the Irish craftsmen John and Hugh Findlay, who had established their workshops in Baltimore.

Duncan Phyfe (1768–1854), like the Findlay brothers, had been born abroad, although he was trained as a joiner and cabinetmaker in America. To many his name is virtually synonymous with the English Regency and American Federal styles of furniture, and the Greek lyre is almost considered his personal motif. More than other

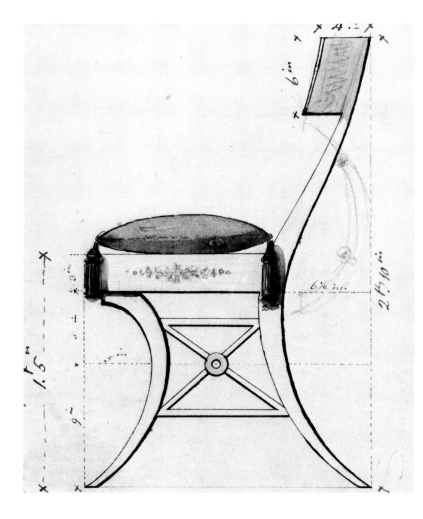

Watercolor drawing by Benjamin Latrobe for the set of chairs that he "designed" in 1809 for the Blue Room of the White House in Washington. They were based on the Greek Klysmos shape shown in Thomas Hope's 1807 book of furniture design and decoration. Maryland Historical Society.

The Findlay brothers of Baltimore probably built this side chair, one of a set of nine. The more elaborate Empire version of the classical style was popular in the new republic. The saber leg of the Klysmos chair has been replaced by the more solid turned Roman legs. Metropolitan Museum of Art. Gifts of Mrs. Paul Moore.

This side chair, one of an original set of 24, displays the characteristic Duncan Phyfe lyre back and reeded side frames. The classical feet were a popular stylistic mannerism. The generally square form and flat sides suggest an accommodation of design to manufacturing convenience. Metropolitan Museum of Art. Gift of the family of Mr. and Mrs. Andrew Varick Stout, in their memory.

craftsmen of his time, Phyfe organized his workshops into a factory by simplifying his designs for volume manufacture. The high quality of his product, however, made him a wealthy man. Phyfe was one of the first Americans to bring his talents as a designer and manufacturer together to advantage.

Most architects, as well as carpenters, masons and builders depended upon imported handbooks to provide basic designs, plans and elevations, and details for replicating one style or another. However, Americans such as George Biddle, the Philadelphia architect, and Asher Benjamin (1773–1845), the Boston architect and teacher of architecture, also began to publish their own books of directions and designs for builders. Asher Benjamin's first book, the *Country Builder's Assistant* (published when he was 24 years old), was so successful that, with 6 other books by him that followed, it ran to more than 40 editions. These books did not slavishly copy their imported models, but rather adapted them to American conditions, allowing for the substitution of wood for the original stone and for the more limited skills and more carefully allocated time and energy of the workmen. Asher Benjamin had a particularly profound influence on American builders in the northeastern United States, where handsome churches and homes grace many early-nineteenth-century villages and towns. Most of Benjamin's designs were Georgian in character, although later he advocated the Greek Revival style as being appropriate to the republican needs of America.

When one style of architecture begins to fade in value, another emerges to take its place by presuming to contain more integrity and to speak its cultural truth more clearly than its predecessor. Even while the Greek Revival style was still reaching its zenith of popularity, a new romanticism in design style was emerging that acclaimed the Gothic as less imperialistic and, therefore, more appropriate to a middle-class society. Today, although the classical style as such has disappeared from public buildings, its cool order still dominates them and its Georgian precedents still predominate in the preference for the Colonial style in American homes and furnishings.

*Front elevation and floor
plans for a country church
by Asher Benjamin. These
provided rural carpenter-
builders with information
for construction in the
popular combination of
Georgian and Federal
styles. Asher Benjamin,
The American Builder's
Companion.*

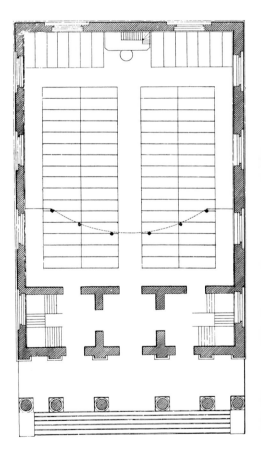

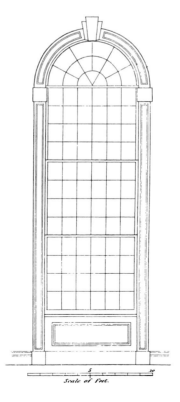

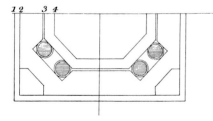

Side elevation for Benjamin's church, with a section of the tower and a detailed elevation of the windows. The use of such guides across the young republic provided aesthetic consistency. Asher Benjamin, The American Builder's Companion.

Scale of Feet.

The Promise
of Free Enterprise

There is in the genius of the people of this country a peculiar aptitude for mechanical improvements. . . . it would operate as a forcible reason for giving opportunities to the exercise of that species of talent by the propagation of manufactures.

Alexander Hamilton, 1791 (84, 21)

The citizens of the young nation were anxious to take advantage of the economic freedom that their struggle for political independence had promised. As heirs to the concept of *laissez-faire,* originally propounded by French social philosophers and later transformed into the principle of freedom of trade by Adam Smith in *The Wealth of Nations,* they eagerly embraced the concept of individual enterprise as one of the rewards of a free society. The Americans found it easy to associate the pursuit of happiness with the acquisition of property, and thus the solemn oath of the Constitution established a unique economy in which the puritan ethic of individual enterprise and economy of means could generate the capital upon which an industrial state could be constructed. (A century later, Arnold Toynbee would confirm Adam Smith's respect for barter as the connection between personal enterprise and a healthy economy: "If we once grant the principle of division of labor, then it follows that one man can live by finding out what other men want. . . ." (87, 56))

Thus, competition between men, each seeking to serve the needs of others, emerged as the most logical means of mutual survival and established in the young American mind a practical as well as philosophic rationale for creating products that would vie with one another in an open market to satisfy consumer needs.

Adam Smith declared that it is in the best interest of a proud and responsible producer to provide those who have faith in him with products of the best quality, and that the consumer is most certain of being served well when he is intelligent and well-informed enough to select and value the better product. Thus, free enterprise may be described as dependent upon an equitable pact between a responsible manufacturer and a knowledgeable consumer. Experience has occa-

sionally proved this to be a puristic theory because productive and promotional energy may, under the press of competition, be misdirected toward those elements of a product that emphasize its apparent quality at the expense of its real quality. And the public, sometimes naively willing to let appearances speak for quality, may prefer to reward promise over performance with its purchase. Nevertheless, the principle of free enterprise has played an indispensible role in the economic development of the United States, resulting in a high standard of living for its citizens.

In the early years of the American republic, many Europeans were convinced that its future would depend upon the development of an agrarian economy drawing upon the seemingly endless natural resources and virgin farmlands. They believed that this economy would be supported by thriving household industries in which collective industrialization would be of little value, agreeing in general with Adam Smith's admonition that "were the Americans, either by combination or any other sort of violence, to stop the importation of European manufactures, and by thus giving a monopoly to such of their own countrymen as could manufacture the live goods, divert any considerable part of their capital into this employment, they would retard instead of accelerating the further increase of their annual produce, and would obstruct instead of promoting the progress of their country." (81, 347)

Smith did not conceive that the vast territory of America would encourage regional specialization that would seek to balance agrarian with industrial growth. He noted that capital earned by manufacturers was used by them to buy uncultivated land for exploitation, and concluded that the advantage did not lie with the "artificer (who) is the servant of his customers, from whom he derives his subsistence" but rather with the "planter who cultivates his own land and derives his necessary subsistence from the labour of his own family (and) is really a master and independent of all the world." (81, 359) It was generally agreed that the development of American industry would be no threat to other countries because its inexhaustible resources would serve as an economic inducement far beyond that promised by its manufactures. The premise and perhaps the intent of the Europeans was that the Americans would be convinced that independence would be best ensured by sending raw materials abroad to be processed by their superiors and then resold to them at a higher price.

These observations were not without justification. The amount of land that was available for clearing, cultivation, and settling exceeded the demand and capacity of the population of the United States, so the promise of a quick profit from the wilderness provided an inducement beyond the risk of investing capital in manufactures. Moreover, with nine out of ten persons living in rural areas, the people were too scattered to provide the concentration of labor that is essential for industry. In addition, there was as yet a shortage of machines and of the technical knowledge to build and operate them.

The leaders of the young republic were therefore divided between those who wanted the United States to remain an edenic paradise and those who believed that economic stability and political independence would depend upon a balance of agrarian and industrial enterprise. John Adams was not at all confident in the ability of the Americans to manufacture. "I say," he wrote to Benjamin Franklin in 1780, "that America will not make manufactures enough for her own consumption these thousand years." (52, 148) Thomas Jefferson was an outspoken champion of those who held to a pastoral ideal as the proper course for the new nation. In his *Notes on Virginia* (1785), he expressed the hope that America would become the home of an agrarian "Golden Age." Jefferson warned that the chief danger to this ideal was the development of machinery, and advised his countrymen to return to their former balance of trade with European industry by exchanging their natural resources and farm products for finished manufactured products from abroad. Jefferson was convinced that the Europeans were obliged to manufacture and export because there was insufficient land for agriculture to sustain them. Therefore he proposed "for the general operations of manufacture, let our work-

shops remain in Europe," noting that "it is better to carry provisions and materials to workmen there, than bring them to provisions and materials, and with them their manners and principles." (34, VIII, 405) Jefferson dreamed of a rural republic built upon the honest labor of the free farmer in the sun rather than the synthetic drudgery of the captive worker in the factory. In 1786, though he had returned from England to report in glowing terms on the progress of the British and his own passion for utilitarian improvements and labor-saving products, Jefferson was still convinced that the future of the United States must depend upon an agrarian economy supported by household manufactures. However, after the War of 1812 with the British, in which American industry played an important part, Jefferson was obliged to come full circle and acknowledge that, although he still preferred the rural life, he realized that "to be independent for the comforts of life, we must fabricate them ourselves," and that "we must now place the manufacturer by the side of the agriculturalist." (32, 745)

The transformation of manufacturing in the United States from a scattered and erratic system of home manufactories into industries coincides to a remarkable degree with the end of the Revolution and the adoption of the Constitution. The Constitution's unified approach to commerce and trade regulations created an atmosphere for equal enterprise that had been all but forgotten in the first scramble for economic advantage by the liberated but not yet joined colonies. At first each state attempted to set its own trade regulations, with the result that other nations (notably England) were able to play one colony against another and to set up trade rules of their own with no fear of concerted American retaliation. In addition, their individual issuance of paper money that proved to be irredeemable had made a shambles out of the fiscal policy of the colonies. However, now, under a common Constitution by which the colonies surrendered their sovereignty to become member states in a republic, a new sense of common economic adventure began to emerge. The passionate will for the success of the union made every citizen aware of his commitment to the whole, and in turn the government acknowledged its dependence upon the ingenuity and industry of its citizens.

Until the Revolution, political and economic circumstances in the colonies did not encourage collaborative industry. Each freeman looked to his own security by way of independent farming, home manufactures, or artisanal activity. However, the spirit that led to revolution and independence now began to stimulate investment in common enterprises such as manufacturing essential commodities in quantities larger than the sum of individual efforts could have produced. The accumulation of capital made it possible for shareholders to invest in the development or acquisition of machines for manufacture (by open or suspect means), to build water mills or steam engines to provide power, and to employ immigrants, itinerant laborers, children, and off-season farmers to operate the machines.

At this point the first signs of what would later be known as the American System of Manufactures begin to emerge. With the developing distinction between the personalized products of home industry and craftsmen and the depersonalized manufactures of industry, a widening gap appears between the managers and the laborers of industry. The managers, as representatives of shareholders who had invested their capital in the enterprise, were obligated to use labor and machines as elements in the manufacturing process; the workmen, on the other hand, were drawn together in order to demand the best possible return for their energy and skill and to guarantee their security. The earliest evidence of industrial corporations in America appeared in the form of societies of private citizens who had invested their capital in the development of manufactures in the interest of profit and patriotism. The Americans were not the first to form such organizations for the promotion of industry, but the concept was entirely in harmony with the search for economic freedom that underlined their conflict with and eventual separation from England.

As early as 1775 the pioneering venture of the United Company of Philadelphia for Promoting American Manufactures had captured the imagination of 20-year-old Tench Coxe, who purchased one share of stock in the company at 10 pounds sterling. Although the company did not succeed, Coxe was moved enough by the potential for such enterprise to write to Benjamin Franklin that American manufactures would result in "considerable *immediate* and immense *future* advantages." Young Coxe, although he was an unpopular figure because of his early sympathies for the British, made himself the champion of a rudimentary American industry. He was convinced that, important as agriculture was, the political independence and survival of the United States depended upon a balanced economy in which manufacturing and agriculture would complement one another. Coxe's pragmatism made him an important force in leading America away from an agrarian society toward a centralized government with economic as well as political power. According to Leo Marx, he was uniquely fitted to his role by a "rare empirical bent which led him to make predictions based chiefly upon economic data collected and interpreted by himself" and a "master publicist's knack of casting aims in the idiom of the dominant ideology." (63, 151)

In May 1787, three days before the opening of the Constitutional Convention in Philadelphia, Coxe was invited by Benjamin Franklin to address the Society for Political Enquiries, an association of fifty leading Philadelphians who met every two weeks to discuss issues of topical interest. His speech, later published as an essay and distributed at the convention, proposed an economic system whereby the cultivators of the soil would provide sustenance not only for themselves and their families, but also for those engaged in manufactures and merchandising, with the surplus to be transported to the best foreign markets:

On one side we should see our manufacturers encouraging the tillers of the earth by the consumption and enjoyment of the fruits of their labours, and supplying them and the rest of their fellow citizens with . . . the necessaries and conveniences of life. . . . Commerce, *on the other hand,* attentive to general interests, would . . . range through foreign climates in search of those supplies, which the manufacturer could not furnish but at too high price, or which nature has not given us at home, in return for the surplus of those stores, that had been drawn from the ocean or produced by the earth. (23, III, 24)

Later in 1787 Tench Coxe was invited by Benjamin Rush to express his views on an American system for balancing agriculture and manufacturing in an inaugural address before an assembly at the University of Pennsylvania of Philadelphians who were interested in organizing a society for the encouragement of manufactures and the useful arts. He called attention to the need to establish manufactures to serve internal markets and ensure a "certainty of supplies in the time of war." Coxe was convinced that the scarcity and high price of labor could be overcome by the encouragement of immigration by those who were tired of conditions in their own country and would come to America in search of freedom and a livelihood through the use of "water-mills, wind-mills, fire [steam power], horses, and machines ingeniously contrived." In response to objections based on the lack of machines in America, Coxe suggested euphemistically that some could be "borrowed" from other nations, and also that American inventors could be stimulated to develop others with offers of premiums to be paid in land. Coxe rejected the conventional concept of the incompatibility of organic and inorganic products, since both were subject to the ultimate laws of nature and the equally inevitable results of man's effort to survive in a hostile environment. Leo Marx believes that the speeches of Coxe "prefigure the emergence of the machine as an American cultural symbol, that is, a token of meaning and value recognized by a large part of the population." (63, 163)

After Coxe's address the assembly voted to establish the Pennsylvania Society for the Encouragement of Manufactures and the Useful Arts, which as one of its first ventures offered a $20 gold medal for the most useful engine operated by water, fire, or any other means that would reduce the labor of manufacturing cloth. In addition, the society attempted repeatedly

(without success) to acquire the knowledge and the machines for spinning fiber and weaving cloth that had been developed in England and whose export was forbidden by the crown. Nevertheless, Coxe continued through 1789 his attempts to acquire, with Jefferson's help, as he wrote to James Madison, "sketches and models from a country in the vicinity of France, which would very much assist the manufactures of the United States."

A few months later Coxe again wrote to Madison about the great importance of machinery: "To procure and record the drawings and descriptions of machinery and apparatus in the arts and philosophical science appears to me a very great object. It is manifest that without depending inconveniently upon manual labor we may, by mechanism and a knowledge of the value of sensible objects and their effects upon each other, save great sums of money, raise our character as an intelligent nation, and increase the comforts of human life and the most pure and dignified enjoyments of the mind of man. No man has a higher confidence than I, in the talents of my countrymen and their ability to attain these things by their native strength of mind." (44, XIII, 112) In this letter Coxe managed to close the triangle of industrial enterprise that characterizes American industry. First he called attention to the promise of profit to be had by saving "great sums of money" in manufacture. Then, patriotically, he promised that the higher level of industry would raise our own "character as an intelligent nation." And finally, he promised that such public service would "increase the comforts of human life." In other words, Coxe saw beyond the simple concept of industry as an instrument for survival. He sensed that it was to become an example of the American ethic whereby personal gain and duty to country could best be achieved by serving public need.

George Washington, in his annual message to the Congress in 1790, observed that the country now had to look to developing independence in manufacturing, particularly insofar as the necessities of defense were concerned: "A free people ought not only to be armed, but disciplined; to which end, a uniform and well-digested plan is requisite; and their safety and interest require that they should promote such manufactures, as tend to render them independent of others for essential particularly for military supplies." (47, 2) Acting upon Washington's recommendation, Congress asked Alexander Hamilton, then secretary of the treasury, to report to it the present state and future possibilities for the development of American manufactures. Hamilton, in his subsequent "Report on the Subject of Manufactures," presented to the House of Representatives in 1791, outlined (with considerable help from Tench Coxe) the economic themes that have since served as a philosophic base for the American balance of agricultural and industrial enterprises. His report countered the prevalent resistance to the encouragement of manufactures; instead it demonstrated a precise knowledge of their indispensibility to the United States without dismissing the importance of agricultural products. Acknowledging the then-popular belief that the United States, with vast tracts of uninhabited territory, did not need to direct energy into the development of manufactures that did not promise the immediate rewards that conquering the wilderness did, Hamilton said that it was reasonable to believe that by specializing in the products of the wilderness and agriculture, the sparse population would be assured of both the goods needed for survival and a surplus that could be exchanged for essential commodities that other countries were in a better position to offer. Hamilton acknowledged that success in the wilderness and in agriculture depended upon dispersal of the population, but pointed out that machinery powered by steam and water was making it possible for industries to operate in or near the growing towns in America.

Although regulations imposed by the English on the Colonies before the Revolution had obstructed the growth of their trade with other countries, Hamilton expressed faith that the United States had not only the immediate capacity to meet their own needs but also a latent ability to generate profitable trade with other countries if they were assured of an international system of fair and free exchange. However, said Hamilton:

The prevalent [policy] has been regulated by an opposite spirit. The consequence of it is that the United States are, to a certain extent, in a situation of a country precluded from foreign commerce. They can indeed, without difficulty, obtain from abroad the manufactured supplies of which they are in want, but they experience numerous and very injurious impediments to the emission and vent of their own commodities. . . . Several countries . . . throw serious obstructions in the way of the principal staples of the United States. . . . The want of reciprocity . . . could not but expose them to a state of impoverishment compared with the opulence to which their political and natural advantages authorize them to aspire. (84, 27)

Therefore, although Hamilton did not condemn the countries of Europe for looking to their own interest, he concluded that, if Europe would not take American products of the soil fairly, the states had no choice but to reduce their dependence upon others for manufactured products.

Since he did not feel that the interests of the proponents of agriculture were in conflict with those of industry, Hamilton proposed that agriculture would be well served by the encouragement of manufacturing. Even if labor were to be directed from the farm to the factory, reducing the amount of land being tilled, there would be an increased demand for farm products. Their value would be raised, and farmers would be encouraged to improve their methods of work in order to produce more. And, in return, a stimulus would be given to those industries serving the farmer to provide him with more efficient tools and machinery. Proof of this is found in the flood of agricultural machines that were invented and developed in the United States in the nineteenth century and in the preeminence of American "agribusiness" in the present century.

Hamilton's report to the House questioned the wisdom of governmental interference in the free choice of opportunity of any one citizen even when it was acting in the best interest of all of the inhabitants of the country. Government's objective, Hamilton proposed, should be to encourage independent initiative by making public funds available when private capital is scarce and by supporting the efforts of its citizens in trade with other countries by way of tariffs, bounties, and premiums. This expression of faith in individual enterprise as a necessary base for communal prosperity is fundamental to the American concept of free enterprise. Hamilton believed that every citizen had a personal right and, in a way, a public obligation to move freely to any location or occupation that held out a better opportunity for him. In this way a diversity of occupations would be encouraged and the spirit of free enterprise would be expanded by the proliferation of those occupations that would support the security and wealth of the nation.

In response to the shortage of enough labor to serve both agriculture and labor, Hamilton suggested that the labor force could be increased by employing persons during their leisure time or during those seasons when they were free of other work and by using women, children, and people who would otherwise be idle because of some physical or mental infirmity. He noted, for example, that in the nail and spike industry in the United States, after the most laborious operations have been performed by water wheels, "of the persons afterwards employed, a great proportion are boys, whose early habits of industry are of importance to the community, to the present support of their families, and to their own future comfort." (84, 80)

Further, Hamilton knew that men will even change nationality if they are offered obvious advantages elsewhere, and he therefore recommended that immigration be promoted by making opportunities to work in manufacturing in the United States attractive to the citizens of other countries. The wealth and industry of the country had already benefited enormously from the contributions of expatriates from Europe. Hamilton suggested that the disturbances in Europe were "inclining its citizens to emigration," so that "the requisite workmen will be more easily acquired than at another time." (84, 61)

In closing his report to the House of Representatives, Alexander Hamilton reiterated the government's commitment to promoting manufactures by stimulating invention by offering "pecuniary rewards according to the occasion and utility" of the contribution and through the acquisition of useful machinery and improvements to machinery from other countries. He noted the value of private organizations for the development of industry, and yet he urged the open support of government where needed because "In countries where there is great private wealth, much may be effected by the voluntary contributions of patriotic individuals, but in a community situated like that of the United States, the public purse must supply the deficiency of private resource." (84, 107)

Hamilton had explored the manufacturing potential of the United States from several angles: the capacity of the country to furnish appropriate new materials, the degree to which the nature of a manufacture admits of a substitute for manual labor in machines, the facility of execution, the extensiveness of uses to which the article can be applied, and its subservience to other interests (particularly those affecting the national defense). Though some manufacturers claimed that the government's involvement in industry was showing preferential interest, most were pleased with Hamilton's report. That the young government came out strongly in favor of private initiative in developing manufactures and offered federal action and monies if needed to support free enterprise did much to set American industry in motion and provided momentum that has carried it forward ever since.

By the time of the American Revolution the English had already put into service machinery invented by Arkwright, Cartwright, Hargreaves, Crompton, and others for the carding, spinning, and weaving of cotton—despite protests from labor, attacks by workers on the machines, and threats against the lives of the inventors. England recognized the economic value of such machinery and processes and passed an Act of Parliament in 1774 establishing strict regulations against their export to other countries. She also forbade knowledgeable artisans from emigrating to other countries. This did not dissuade the Americans from attempting to pry the secrets of textile manufacturing from England.

The credit for building and operating the first successful cotton mill in the United States is generally given to Samuel Slater, who had emigrated from England in 1789, when he was 22 years old, with papers showing that he had been apprenticed to a partner of Arkwright, the inventor of the first spinning machine. Slater was apparently enticed by an advertisement placed by the Pennsylvania Legislature that a reward of 100 pounds would be paid for a new carding mill and the establishment of a society to promote cotton manufactures. In the United States, Slater signed an agreement with Almy and Brown of Providence, Rhode Island, to build a series of Arkwright machines. However, since he had left London without drawings or models in order to avoid suspicion, Slater was obliged to draw each part of the machine on oak boards to be cut out and put together with dowels while a blacksmith made the metal parts according to his descriptions. After a series of frustrating attempts, a small manufactory was built in 1793, operated originally by human treadle power. Slater was "the first in America to achieve commercial profitability with Arkwright technology." (50, 84)

Thus, technology and industry came to the United States somewhat clandestinely, as an extension of the great sweep of the industrial revolution in Europe that was beginning to replace the medieval mysteries of the guild system with a more open system of evolution that stimulated rather than inhibited initiative. For centuries, it had been customary for monarchs to grant "privileges" for various trade activities to those who made discoveries that were useful to the state and profitable to the crown. In the United States, by the end of the eighteenth century, to profit from a new discovery had became a right rather than a royal privilege. Even before the Revolution the various colonies had laws conceived to encourage inventors by protecting their right to profit from their genius. As early as 1650 the courts of the Massachusetts colony had passed a law that there "should be no monopolies but of such new inventions as were profitable to the country and that for a short time only." This concept of government supporting new inventions by guaranteeing the inventor a monopoly for a limited period in return for public disclosure of his idea was incorporated into the Constitution of the United States in article I, section 8: "The Congress shall have power to promote the progress of science and useful arts, by securing for limited times to authors and inventors the exclusive rights to their respective writings and discoveries."

The U.S. government began operations in March 1789, and in January 1790 President Washington, addressing the second session of the First Congress, implored it to give "effectual encouragement . . . to the exertion of skill and genius at home." In February the patent bill was presented to Congress and passed by the House and the Senate. It defined a patent as "any useful art, manufacture, engine, machine, or device, or any improvement thereon not before known or used." (The patent act was concerned primarily with the scientific and technological uniqueness and the utility of a product, a process, or a composition; it was to be a half-century before the government would recognize its obligation to protect a unique appearance or design.) At first the term of protection was seven years, and was renewable for another seven. In 1861, the period of protection would be set at 17 years and declared not renewable. Congress vested original responsibility for patents in a board that included Secretary of State Thomas Jefferson, Secretary of War Henry Knox, and Attorney General Edmund Randolf, who were granted the power to issue patents "if they shall deem the invention or discovery sufficiently useful and important." (Although Jefferson never took out a patent, he found great merit in limited monopolies as a means of encouraging the development of "many ingenious improvements . . . in the arts, and especially in the mechanical arts." (88, 462) Jefferson's own inventions included an improvement in the mold board of a plow that was awarded a gold medal from the Agricultural Society of Paris. His zealous interest in the value of new things was an important factor in the initial acceptance and success of the American patent system.)

Three patents were granted in the first year. The first, signed by Washington and Jefferson among

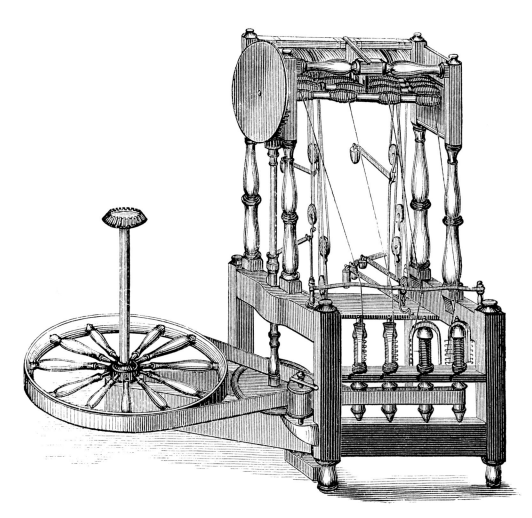

Richard Arkwright's
drawing for a spinning
machine, patented in En-
gland in 1769. The ma-
chine—first driven by
horse power, then by a
water wheel, and finally by
steam—industrialized the
cotton mills and thus
marks the origin of the
factory system. Harper's
Monthly, December 1874.

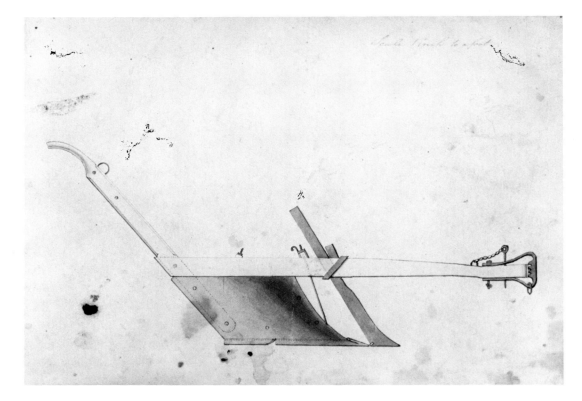

Thomas Jefferson was
fascinated with mechani-
cal devices. Among the
contrivances he invented
but did not patent is this
mold board for a plow
(inspired by the form of
a xylophonelike musical
instrument of the time).
The mold board caught
the turf as it was lifted by
the plow and turned it
over. Courtesy of Massa-
chusetts Historical Soci-
ety; Kovacik photograph.

others, was granted to Samuel Hopkins of Vermont for a new apparatus and process for making potash and pearl ash. In 1800, ten years after the patent law was enacted, the number of American patents had grown to 306—about half the number of British patents. By 1820, American patents had surpassed the British 1,748 to 1,125, and by 1870 American patents granted were double the number of British ones. These figures are representative of the difference in the thrust toward technology between the mother country and her rebellious offspring, and serve to demonstrate the healthy stimulus provided by the Patent Law of the United States.

The opportunities and the promise of rewards for ingenuity and initiative released in the American what Michel Chevalier called "a mechanic in his soul" (20, 285)—an instinct for contriving mechanical substitutes for manpower and for finding the quickest and easiest way of getting things done. This became such an obsession that the typical Yankee mechanic seemed driven to use the leisure he gained with one device to think up another one so that he would have more time to conceive still other improvements.

Oliver Evans (1755–1819), a man driven by the insatiable curiosity that plagues the creative spirit, was one of the first Americans to move into the free space of expanding science and technology. He typifies that rare breed of humans who are predestined—happily for society, but often tragically for themselves—to a life of unremitting and often unrewarding labor in search of a better world to come. Evans reflected this philosophy in his own hand in a copy of his book *The Abortion of the Young Steam Engineer's Guide* that he willed to his son: ". . . he that studies and writes on the improvements of the arts and sciences labours to benefit generations yet unborn, for it is not probable that his contemporaries will pay any attention to him, especially those of his relations, friends and intimates; therefore improvements progress so slowly." (6, IV)

As a boy Evans was fascinated with the potential of deriving power from water and fire, and as an older man he was to devote himself to developing steam engines and applications for them. His first invention, however, was a machine to draw and bend wires for carding cotton and wool. Like Franklin before him, Evans refused to apply for legal protection because he believed that everyone should be able to benefit from his ideas without restriction. His professional career, nevertheless, was marked by conflict with others who sought to deprive him of the rewards of his creativity. At one point he even burned many of his designs and papers in frustration.

Oliver Evans made an important contribution to American industry when he demonstrated that ingenuity could be applied not only to the products of manufacture but to the process of manufacturing itself. In 1785 he built the first automatic factory in America, a grist mill that was capable of performing all of the functions of a flour mill with water power. Evans was granted exclusive privileges to the process for 14 years by the state of Pennsylvania. In order to promote his invention, he wrote and published *The Young Mill-wright and Miller's Guide,* which passed through fifteen editions between 1795 and 1860 and is still used as a reference by millers.

In 1800 Evans built the "Columbian" high-pressure steam engine and put it to use for grinding plaster and cutting stone. In 1804, his dream of using steam for locomotion came true when he was contracted by the Philadelphia Board of Health to construct a steam-operated machine for dredging and cleaning docks. He built his contraption, 12 miles from the Delaware River, on wheels linked to the engine with a belt. After the machine had rolled itself to the water's edge, the belt was shifted to a paddleboard to drive it across the water to its work site. Evans advertised his machine as the *Oruckter Amphibolos* (amphibious digger), and charged the public 25 cents each to view it in order to raise money to help pay the workers who had helped him build it. Although he never achieved his dream of building a locomotive, he quite accurately predicted its inevitability, not as an abstracted visionary but rather from the conviction and experience of a practical mechanic.

Plate XXI.

An elevation section of the automatic grist mill built and operated by Oliver Evans. Evans, The Young Mill-Wright and Miller's Guide.

Cross-section of Evans's automatic grist mill. Evans, The Young Mill-Wright and Miller's Guide.

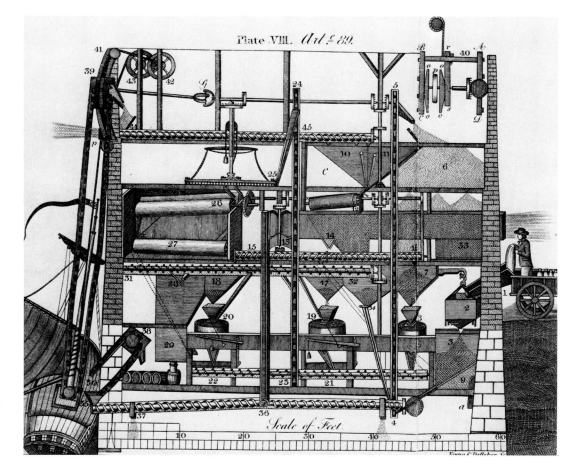

Plate VIII. *Art. 89.*

Scale of Feet

Oliver Evans also adapted
the high-pressure steam
engine that he built in
1802 to a scow for dredg-
ing the Delaware River. It
was the first American vehicle to run on both land
and water. The band could
be connected either to the
wheels or to the stern-
mounted paddlewheel.
Harper's Monthly, De-
cember 1874.

Despite Evan's progress, it was apparent that the first major avenue of mechanical transportation in the United States would have to be the waterways. The steam engine was more easily carried on water than on land, and the basic principle of water propulsion seemed to be more readily accomplished by the then-available technical means. It is all but impossible to determine who, if anyone, should be given the fundamental credit for inventing the steamboat. The mechanical system had been established in England in 1780, and in that same year a Frenchman, the Marquis de Jouffroy, had successfully operated a steamboat. In 1785, Joseph Bramah of England patented a rotary engine with a screw propeller. Three years later, John Fitch (1743–1798) of Philadelphia, a clockmaker, silversmith, and inventor, conceived and built his first steamboat, a rather curious vessel that employed 12 vertical reciprocating paddles. A later version that ran on schedule between Philadelphia and Trenton on the Delaware River was the first successful experiment in steam navigation in Europe or America. Later, disappointed by lack of recognition and failure in other projects, Fitch committed suicide.

While Robert Livingston was in Paris in 1803 on assignment for President Jefferson to negotiate the purchase of the Louisiana Territory, he met the precocious American artist-inventor Robert Fulton (1765–1815) and managed to interest him in designing a steamboat. Fulton had been artistically inclined since childhood, when he had made pencils and taught himself to draw. He was already well known as a painter and inventor when he decided to go to London to paint with Benjamin West, an American expatriate artist who was a friend of his father. In England, Fulton had combined painting with design and had received patents for a cast-iron aqueduct, a method of raising canal boats, and the first power shovel for excavating canals. He had also written a treatise on a new system of canal navigation. In 1797 Fulton crossed the channel to take out French patents for some of his ideas, and was commissioned there to design a submarine. Although it was not the first submarine to be designed by an American, his *Nautilus* was con-

sidered successful. It was not permitted to approach near enough to British ships to hang bombs on them, but it scared them away.

Back in the United States, Fulton accepted a commission from Livingston to build a steamboat. In 1807 his boat, the *Katherine of Clermont,* was launched on the Hudson to win a competition to establish regular passenger service between New York City and Albany. Running on two side paddlewheels driven by an English Boulton and Watt engine, the *Clermont* was capable of making the voyage of 150 miles in 33 hours. Although a committee of the New York legislature declared in a hearing that the *Clermont* was in substance the same as the boat that was patented by John Fitch in 1791, history, with its capacity to put events in neat romantic order, generally credits Fulton as the father of American steam navigation. In 1811, Fulton, again with Livingston, designed and built in Pittsburgh the sternwheeler *New Orleans,* the first steamboat to ply the Ohio and Mississippi between Pittsburgh and New Orleans. His successful steamboats opened the inland waterways and provided a flow of goods that was at least equal in volume to transatlantic shipping.

In response to the expansive mood of the young republic, the wheels of American industry were beginning to gather momentum. Slater's cotton mills increased in number, and they soon were powered by water wheels rather than human-driven treadmills. However, the potential of these mills was not realized immediately because the volume of cotton grown in the United States was barely enough to serve home manufactures. It was evident that industry could not grow until the sources of supply and the methods of manufacture, distribution, and consumption were in dynamic balance. What was needed most was an increase in the supply of cotton fiber. Although cotton was being grown, it was of little commercial value because of the difficulty of separating the seed from the fiber. It took 10 hours of labor to "clean" 3 pounds of seed from a pound of fiber.

Eli Whitney (1765–1825) had recently graduated from Yale University and was serving as a tutor to a private family in Georgia when he learned about the cotton-cleaning problem and was

John Fitch attempted
to use a steam engine to
power a boat with paddles
duplicating human effort.
Although impractical and
weighing 7 tons, it was
able to move 3–4 miles
an hour. J. Franklin
Reigart, The Life of Robert
Fulton.

The Katherine of Clermont,
designed and built by
Robert Fulton, established
steam locomotion on
American rivers by making
the first trip from New York
to Albany on the Hudson.
It was 150 feet long with
an 8-foot beam and a
7-foot hold, displacing 100
tons. Courtesy of New York
Historical Society.

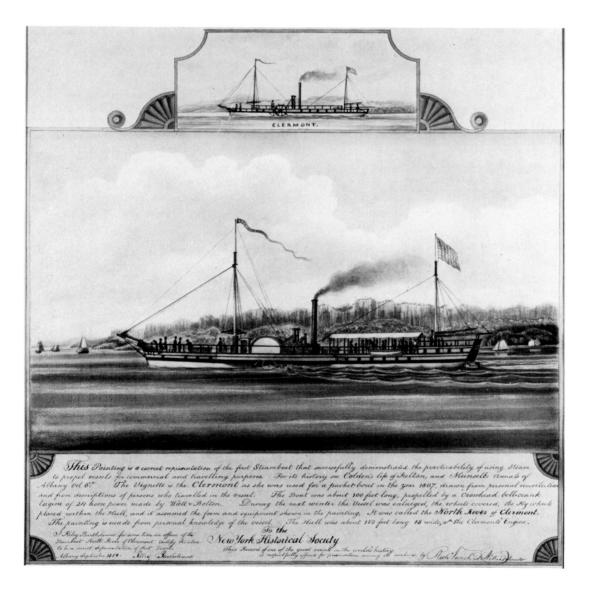

CLERMONT.

This Painting is a correct representation of the first Steamboat that successfully demonstrated the practicability of using Steam to propel vessels for commercial and travelling purposes. For its history see Colden's life of Fulton, and Munsells Annals of Albany Oct 6th. The Vignette is the Clermont as she was used for a packet-boat in the year 1807, drawn from personal recollection and from descriptions of persons who travelled in the vessel. The Boat was about 100 feet long, propelled by a Crosshead bell-crank Engine of 24 horse power made by Wall & Bolton. During the next winter the Vessel was enlarged, the wheels covered, the Fly wheels placed within the Hull, and it assumed the form and equipment shown in the painting, It was called the North River of Clermont. The painting is made from personal knowledge of the vessel. The Hull was about 150 feet long 18 wide, wth the Clermonts' Engine.

To the
New York Historical Society

I Riley Bartholomew for some time an officer of the
Steamboat North River of Clermont certify the above
to be a correct representation of that Vessel.
Albany September 1858. Riley Bartholomew

This Record of one of the great events in the world's history
is respectfully offered for preservation among its archives by Richd Varick DeWitt Director

stimulated to try his own hand at solving it. In a flash of insight, he conceived and built a small working model of a machine to do the job. With a full-size water-powered cotton gin, one man could clean 300–1,000 pounds of cotton a day. This invention closed the circle of operations necessary to make the cotton industry roll. Within two years, cotton production increased eightfold, and a quarter of a century later America was not only meeting her own burgeoning demands but also supplying three-fourths of the cotton used in England.

Before Whitney could patent his invention, its secret was discovered and quickly pressed into service without permission. Other than the relatively small payments made to him in conscience by the states of North and South Carolina, he earned nothing from his invention. He complained of his sad experience in a letter to his fellow inventor Robert Fulton, and returned to New Hampshire determined to turn his inventive genius to other enterprises that would prove more rewarding.

Eli Whitney's brilliant breakthrough to a solution of the problem that had held down the commercial production of cotton in America illustrates that a perceptive outsider may often see a solution more clearly than those on the inside. The "sophisticated naivete" of a designer may capture an abstraction or glimpse an order that escapes a person whose proximity to the problem has obscured his vision. Whitney's cotton gin brought prosperity back to southern plantations, and with it the revival of slavery. And it gave rise to the manufacture of cotton cloth in the North.

Even so, it would be misleading to conclude that home manufactures disappeared overnight. In 1810 most of the cotton and wool fabric in the United States was still being manufactured in the home for personal use and private sale. Home manufactures—for the rural family a profitable source of income that could be gained with little or no loss of farm time—continued to increase for a while.

Standardization was apparently applied first by the French Artillery in 1765 when the various parts of gun carriages were organized so that any part could be replaced by another in any carriage in its class. Interchangeability had also been used by the British in the manufacture of pulley blocks for the Navy. The first American knowledge of part standardization came in 1785 when Thomas Jefferson wrote from France of having observed it in the construction of muskets in the manufactory of LeBlanc. Jefferson wrote of being given the parts of 50 musket locks: "I put several together myself, taking pieces at hazard as they came to hand, and they fitted in the most perfect manner." His efforts to bring LeBlanc to the United States failed because of the threat the new method posed to those using traditional methods of building weaponry.

A few years later, however, the principle of standardization and interchangeability bore fruit when, in 1798, Eli Whitney proposed to Oliver Wolcott, secretary of the treasury under John Adams, that he be contracted to manufacture 10,000–15,000 stands of arms (consisting of musket, bayonet, ramrod, wiper, and screwdriver). Whitney promised to deliver them within two years by manufacturing them on the principle of interchangeability. "I am persuaded," he wrote to Wolcott, "that machinery moved by water, adapted to this business, would greatly diminish the labor and greatly facilitate the manufacture of this article. Machines for forging, rolling, floating, boring, grinding, polishing, etc., may all be made use of to advantage." (205, V, 117) Although Whitney had no factory or machinery, his reputation as the inventor of the cotton gin gained him $30,000 in bonds from New Haven friends and $10,000 from a New Haven bank to win a government contract of $134,000. With these resources he set up a mill at what is now called Whitneyville, Connecticut, and began operation in 1800 with 60 employees. Before Whitney could begin production he had to redesign the musket, anticipate and plan every stage in its manufacture, devise appropriate patterns and processes for each part, and either acquire appropriate machinery for each operation or design and build it where none was available. Like others before him, Whitney found it necessary to invent

Eli Whitney's cotton gin, which paved the way for mass production in the United States by providing cotton fiber to feed the spinning and weaving machines. P. G. Hubert, Men of Achievement.

*Eli Whitney's gun factory
at Whitneyville in 1826.
Behind the factory is a row
of houses that Whitney
built for his employees —
one of the earliest housing
projects in the United
States. Yale University Art
Gallery, Mabel Brady
Garvan Collection.*

machines that would ensure the accuracy that was needed. His contribution was the milling machine, just as before him (in England) John Wilkinson's boring machine had made Watt's steam engine efficient and Henry Maudsley's machine lathe had made the manufacture of other accurate machines possible.

Although Whitney did not meet the deadline of his original contract (it took him nearly 8 years to work the "bugs" out of his system), the order was completed and the interchangeability of his musket components was demonstrated successfully in Washington. As a result he was awarded a new contract in 1811, this time for 15,000 stands, and he managed to fill that order within 2 years. In the heat of the war of 1812, it became evident that the greatest impetus for the transformation of science to technology would come from military needs.

Eli Whitney's musket factory displayed the basic characteristics of what Henry Clay first called the American System of Manufactures—the method by which a successful industry could be built "from the top down" by a group of citizens who shared a perceptive and even inspired sense of public needs and desires and were willing to combine their personal capital and resources in a joint venture that would serve the public while returning a fair profit to them for their initiative. It demonstrated that intelligent planning in combination with empirical experience could be used to design and build under one roof a complete facility with a common source of power; to acquire or develop specialized machines and to link them into a system; to attract, train, and compensate the necessary labor force; to manufacture a product or a range of products in sufficient quantities to meet anticipated or generated demand; and to amortize the costs of building, manufacturing, and marketing by prorating them across the sales of the products. The American System of Manufactures was, according to John Kouwenhoven, "as indigenous to the United States as the husking bee." (55, 40)

The natural corollary to a successful system of manufactures was the development of a dependable system of transportation. George Washington had said in 1785 that improved transportation was necessary to bind the people of the United States together. Alexander Hamilton's interest in transportation was motivated by economics rather than politics. He believed that improvements would unite the remote and the populous parts of the country in a common market, and that the introduction of rival commodities would break down monopolies in areas where they had existed.

The first 50 years of freedom had transformed the new republic from a group of divided colonies into a virile young industrial competitor on all fronts. While the concept of technology as the adaptation of scientific knowledge to useful purpose was as yet unrefined (the word *technology* was coined in 1829 by Jacob Bigelow of Harvard), the citizens of the United States were quickly putting their dreams of a pastoral America into balance with the economic realities of industry. Technological centralization was replacing the vertical structure of craft technology (in which the artisan participated in every stage of his product's evolution) with a horizontal stratification (in which one layer of employees is charged with planning a product, another with its manufacture, and a third with its merchandising). It was at this point that the emergence of the industrial designer as the catalyst between these three layers and the consumer became inevitable.

The American System
of Manufactures

The spirit of manufacture has taken deep root among us and its foundations are laid in too great expence to be abandoned.

Thomas Jefferson, 1809 (62, 124)

By 1810, on the eve of the second war with England, the second census of the United States estimated the value of manufactures at $127 million. This was revised by Tench Coxe to at least $172 million, exclusive of agricultural and natural products—better than double the figure for 1790. By this time, as reported by Albert Gallatin, secretary of the treasury under Jefferson, the United States was exporting more than it was importing of many products, and thus could be considered self-sufficient in those areas of manufacturing.

This growing manufacturing prosperity led to increasingly bitter rivalry between the former colonies and England. When England, in her war with France, blockaded the French coast and seized American merchant ships that were attempting to trade with the French, she refused to recognize naturalized American sailors and impressed most of them into naval service. At first Jefferson retaliated by stopping all trade with Europe. This nearly ruined American shipping before the new president, James Madison, changed the regulation to apply only to England. Although the British then lifted their blockade of American ships, word did not reach the United States until after Congress had declared war on England. After a costly exchange of victories and defeats, the conflict was finally brought to a close with the signing of the Treaty of Ghent in 1814.

The unwanted war was not without benefits to the United States. It tested the strength of the young nation, made Andrew Jackson an important political force, and had what appeared to be a beneficial effect on industry and the general economy. By cutting off all foreign imports, it generated a boom in American manufacturing not unlike that to be repeated a little more than 100 years later after World War I. Many stock companies were hastily formed to establish industries of every kind, with the result that native competition for labor and materials pushed wages much higher

as raw materials doubled and tripled in value. A hollow sense of prosperity was also stimulated by the inflationary policy of issuing bank notes and treasury bonds with no backup in hard gold or silver. As a result, money was losing value at the same time that prosperity appeared to be increasing. As usual, the public ignored warnings of impending disaster and gloried in the belief that it had achieved a permanent state of prosperity.

The postwar period, however, brought a drastic drop in the demand of the government for products, and when the strong flow of imports began again the buyer's market that resulted brought on a calamitous drop in prices that ruined many merchants and many of the new manufacturing companies. In 1816, Congress attempted to turn the tide by invoking strong protective tariffs along the lines suggested by Alexander J. Dallas, secretary of the treasury under Madison. Dallas divided American manufactures into three classes that help to define the state of manufactures of the time. In the first class he placed those products that were firmly established and whose production met all or almost all of the demands for domestic consumption. It included cabinetry and woodwork, carriages, weaponry, iron castings, window glass, paper, books, leather products, and hats. For these products Dallas recommended a protective tariff as high as 40 percent. The second class comprised those manufactures whose production was not yet sufficient to meet domestic demands but could be stimulated with the help of a smaller tariff. This class included larger iron manufactures plus shovels, spades, axes, hoes, scythes, nails, pewter, tin, copper and brass manufactures, coarser fabrics, and spirits, beer, ale, and porter. Dallas's third class covered products whose production was yet so limited as to make the United States almost entirely dependent upon foreign sources. It included the finer fabrics, silk, hosiery, gloves, some hardware, cutlery, pins and needles, china, porcelain and earthenware, and glass holloware. Dallas recommended that tariffs in this class be permitted to rise and fall according to changes in domestic manufacture and expressed a strong sentiment in favor of stiffer tariffs in order to help the weaker companies.

The economic depression was too deep to be easily turned upward again, and not until 1819—after the United States Bank resumed species payment and after the prices of commodities, employment, and industrial activity had been reduced to more logical levels—did the general economic state of the nation begin to improve. The citizens of every state began to organize societies for manufacture, such as the American Society for the Encouragement of Domestic Manufacture, founded in 1816 in New York City, which invited manufacturers, merchants, scientists, and men and women everywhere to cooperate in building American industry. The society, with active assistance from Daniel D. Tompkins, vice-president of the United States, established additional groups in Baltimore, Lancaster, Hartford, Middletown, and other cities in New England and the more western states. In 1817, President James Monroe and past presidents Adams and Jefferson were also elected members of the American Society. In the interest of a healthy national economy, it seemed entirely appropriate at the time for elected officers of the country to lend their personal support to the expansion of manufactures. That same year, the Columbian Institute for the Promotion of the Arts and Sciences was incorporated in Washington. By 1823 over 200 manufacturing companies had been issued corporate charters by New York State, and hundreds of unincorporated companies were also engaged in manufacturing there. By the end of the decade a second New York organization, the American Institute of the City of New York, was established for the purpose of encouraging and promoting domestic industry by conducting exhibitions of machinery and manufactures and maintaining a library and a collection of models.

The New England Society for the Promotion of Manufactures and the Mechanic Arts was organized in 1826 in Boston, and within five years there were some 240 manufacturing concerns active in Massachusetts. In a pattern similar to that of the Conservatoire des Arts et Métiers of Paris and the National Repository in London, the New England Society began to hold exhibitions, and to award prizes to new and useful inventions, machinery, experiments in chemistry, and natural philosophy and premiums for the best

specimens of the skill and ingenuity of mechanics, in the halls over Faneuil Market in Boston. The first five sales of industrially manufactured products brought $2 million.

These combines of private citizens uniting in purpose and joining their resources for personal gain and for the economic stability of the nation were the forerunners of the institutions of democratized capital that characterize American industry. It was natural, therefore, that an interest in the art and science of design and engineering should emerge as men found new meanings for words such as *machine, engine, mechanics,* and *manufacture* and learned to marshall their experience into systems of knowledge upon which more specialized careers could be built. Information was organized into subject areas for the lecture halls of the institutes, and young men sought out their classrooms as springboards for careers in industry. The concept of design as a means by which a plan for a product could be conceived in the mind and laid out in detail for analysis and evaluation before it was manufactured began to expand the traditional use of the word *design* beyond artistic composition and decoration.

William Dunlap, in his monumental *History of the Rise and Progress of the Arts of Design in the United States* (1834), describes the contemporary use of *design:*

Design, in its broadest signification, is the plan of the whole, whether applied to building, modeling, painting, engraving, or landscape gardening; in its limited sense it denotes merely drawing; the art of representing form. (28, I, 9)

The progress of the arts in design is from those that are necessary to those that delight, ennoble, refine. Man first seeks shelter from the elements, and defense from savages of his own, or the brute kind. In his progress to that perfection destined for him, by his bountiful Creator, he feels the necessity of refinement and beauty. In this progress, architecture is first in order, sculpture second, painting third, and engraving follows to perpetuate by diffusing the forms invented by her sisters. (28, I, 10)

The mechanic arts have accompanied and assisted the fine arts in every step of their progress. To the sciences, they have been indispensable handmaids. In all the ameliorations of man's earthly sojourn, the mechanic and fine arts have gone hand in hand. The painter, the sculptor, the engraver, and the architect, will all acknowledge their obligations to the mechanic arts, and the mechanic will be pleased by the consciousness that he has aided the arts of design in arriving at their present state of perfection. (28, I, 11)

The line between artist, the designer, and the mechanic was not as clearly defined then as it seems to be now. As noted above, Robert Fulton was an artist before he abandoned painting to apply his knowledge of drawing to illustrating the inventions for which he is best remembered. At about the same time, Paul Revere left engraving and silversmithing to become a manufacturer of church bells, brass cannon, and copper products. In 1795 Revere was elected the first president of the Massachusetts Charitable Mechanics Association. Charles Willson Peale (1741–1827) was at various times a saddler, a coachmaker, a clock- and watchmaker, a silversmith, and a portrait painter. He established a museum of natural history and gave lectures on natural philosophy, and in 1794 attempted to establish a school in Philadelphia for the Art of Designing. And Samuel F. B. Morse (1791–1872) was already a distinguished painter before he invented the telegraph. In 1826 Morse was elected president of the National Academy of Design, an organization that still exists.

Interest in the sciences went beyond the range of earlier societies that had been founded by intellectuals for the exploration of natural phenomena. Such organizations as the American Philosophical Society (founded by Franklin in Philadelphia in 1743), the American Academy of Arts and Sciences (Boston, 1780), the Cincinnati Academy of Arts and Sciences (1799), and the Literature and Philosophical Society of New York (1814) recognized that knowledge acquired by scientific research and experiment could be applied with benefit to the development of manufactures. Furthermore, since the amount of accumulated knowledge had grown too great to be transmitted by apprenticeship, it was inevitable that institutes of specialized education would have to be organized in the United States.

Such institutes were also being established abroad. France had opened the Ecole Polytechnique late in the eighteenth century as a collection of departments of specialized learning. Its founder G. Monge produced the first systematic scientific treatise in 1799, and in 1811 his associate J. P. N. Hachette was the first to apply the methods of solid geometry to the construction of machinery. The first such institute in the British Isles, which was organized in 1800 in Glasgow after a series of lectures by George Birkbeck, was reconstituted in 1824 as the Glasgow Mechanic's Institute. By 1821 another similar institute had been established as the Edinburgh School of Arts, and it was followed by others at Aberdeen, Newcastle, Manchester, and Liverpool. The first organization to attract popular attention to this important movement in the applied sciences and technology was the London Mechanic's Institute, established in 1823 at the suggestion of the Mechanic's Magazine of London for the promotion of useful knowledge among the working classes. Within a year, 23 other institutes had been established in the British Isles. Sir Humphrey Davy, the English natural philosopher, expressed around 1820 the state of mind of many of his contemporaries: "The love of knowledge is a faculty belonging to the human mind in every state of society: and it is one by whom it is most justly characterized—one the most worthy of being cultivated and extended." (80, 3) Germany was already operating trade schools by 1817. The first Polytechnic school, patterned after the Paris school, opened in 1825 at Karlsruhe. By the end of the 1820s there were other schools in Munich, Dresden, Stuttgart, and Hanover. Klemm describes them as "the nurseries of a scientifically cultivated technology which vigorously advanced the industrial movement supported by middle class culture." (54, 317)

All of the institutes offered series of lectures on the principles of science and the useful arts. Tickets were sold for sessions that met two to four times a week over a period of several weeks. In addition, the institutes maintained scientific libraries and collections of models of machinery, "philosophical" apparatus, and minerals and other curios of natural history. These mechanic's institutes were the forerunners of the polytechnics of Europe and the institutes of technology that exist today in the United States on an equal level with the universities.

The first American mechanics institutes were founded in 1824 in New York and Philadelphia, and shortly afterwards others opened in Boston, Baltimore, and Troy. These schools were primarily for men, although the Baltimore school offered a course in straw plaiting for young ladies and the New York school opened a school for women in 1830. The New York Mechanics Institute stated its fundamental purpose (somewhat effusively) as "shedding the light of useful science over the paths of the mechanician [and] pouring upon the mind of the young artisan the previous truths of accumulated knowledge." (90, 5) Its program included a popular series of lectures in modeling, architectural and ornamental drawing, and machinery.

The second and perhaps the most successful institute in the United States was the Franklin Institute of the State of Pennsylvania, in Philadelphia, whose membership was open to the public for $2. Its objectives and operations were quite similar to those of the New York Institute, except that from its outset the Franklin Institute elected to draw no lines separating the various trades and professions. The design and construction of buildings, machinery, and domestic wares and appliances all found a common ground in a search for a common language and similar tools of research and communication. Before the end of the Institute's first year, its successful lectures had made possible professorships in natural philosophy, chemistry, and mineralogy. In October of 1826 the Franklin Institute held in Carpenters Hall the first exhibition of the products of American industry at which gold, silver, and bronze medals were awarded for outstanding work. It was hoped that "artists and manufacturers of every kind, anxious to display their workmanship, would be willing to pay for the privilege." (90, 27) Exhibitions were held annually for some 30 years, and occasionally thereafter, with their main purpose stated as "to encourage and stimulate the industry and ingenuity of American artisans and manufacturers; to introduce customers and producers to one another; to acquaint our merchants with the elements and materials of commerce; and our statesmen and the public with the resources of the country." (130, 5)

In 1826 the Franklin Institute offered premiums for, among other products, table knives and forks, flint glassware, a scale beam, a roller for a silversmith, and a smith's anvil. Another premium was offered for the best-constructed grate, or stove, for burning anthracite coal. The announcement for the grate competition stated that "tastefulness of design, though not a primary object, will be considered, as far as it is compatible with economy." Another premium offered for the best cabinet secretary and bookcase stated that "regard will be had, in awarding the premium on cabinet ware, to the taste exhibited in the design, as well as to excellence in workmanship." (130, 7) As a result of the stimulus provided by such exhibitions and competitions, new products appeared all along the frontiers of technology. The requirements stated for these premiums bracket the contemporary profession of industrial design as concerned with the practical balance of function and economy with convenience and aesthetics.

Thomas P. Jones, the first editor of the *Franklin Journal and American Mechanics Magazine,* commented somewhat sarcastically in the second issue (February 1826) about the change in status of the arts and sciences: "[They] did not at first attract, as they deserved, the attention of the wealthy and the noble. By them, education was, for a while, despised, as alone becoming the inferior classes of society; and it was deemed the distinctive badge of opulence and high birth, to be among the most ignorant in the land." (131, 66) Jones went on to advise his readers that the acquisition of knowledge in the mechanical arts could also lead to power and wealth.

The Franklin Institute considered its two branches of drawing very important to the development of designers and mechanics. One drawing course was essentially freehand drawing as related to the fine arts, including instruction in useful and ornamental drawing. The other course, in architectural drawing, was said to comprise "all the departments of mechanical drawing . . . connected with various occupations" and to be "necessary to Cabinet-makers, Carpenters, Stone-cutters, and mechanists" so that "they, by their being instructed in a course of geometrical drawing, may acquire a knowledge of designing, relative to their professions, upon sound principles." (132, 190) The two methods of drawing were conveniently brought together in 1829, when the *Journal* described the introduction by Professor Farish of Cambridge, England, of a "new mode of drawing in perspective which is peculiarly applicable to the delineation of machinery, as the proximate and most distant parts of the thing represented which lie in the same plane, are all drawn to the same scale." The designation *isometrical* was said to have been given to the method by its inventor. (133, 429) A number of years later, the Franklin Institute was still concerned about drawing. At the close of the thirteenth annual exhibition, the *Franklin Journal* observed "with what awkwardness and difficulty do they express their ideas of machines, buildings, furniture and a thousand other objects, merely because they cannot draw." (138, 38)

The popularity of the courses offered by institutes in various cities of the young republic are indicative of the importance the average American placed upon practical education as an avenue to success. "A science is taken up as a matter of business," wrote Tocqueville, "and the only branch of it which is attended to is such as admits of an immediate practical application." (86, I, 44) Men now aspired to make their fortunes either by inventing or developing new products for manufacture or by joining one of the growing American industries. In a way, a frontier as dramatic as the West had been opened to Americans by the heady promises of industrialization and the consolidation of the country's economic position in the community of nations.

In the early eighteenth century, as today, manufacturing enterprises were chartered by government and brought into corporate being by the investors. The shareholders then turned responsibility for the development and direction of the venture over to managers. Success depended on management's sensitivity to the public's needs and desires, upon the ingenuity with which they conceived or acquired machinery and organized and operated the manufacturing process, and upon their ability to put together an effective system of distribution and marketing.

The workers in the manufacturing enterprises included immigrants, itinerant laborers, off-season farmers, men, women and even children. Though the end goals of managers and workers were similar, their obligations to industry seemed at times to be antithetical. As Tocqueville wrote: "Whereas the workman concentrates his faculties more and more upon the study of a single detail, the master surveys a more extensive whole, and the mind of the latter is enlarged in proportion as that of the former is narrowed. In a short time, the one will require nothing but physical strength without intelligence; the other stands in need of science, and almost of genius, to ensure success." (86, II, 191)

As the contrast between master and workman became more distinct, a new form of aristocracy began to spring up. However, the artistocracy of management did not for the most part possess the spirit of benevolent paternalism that characterized the landed aristocracy of earlier times. Whereas urban and rural artisans had had individual control over their methods of work and the quality of their products, now workers were transformed into laborers—essentially, human elements in a manufacturing process, obliged to contain their physical and mental capacities within strict limits imposed by management. The heart of the factory system was found in the contrapuntal beat of management and labor relations.

Factory managers were obliged to offer high wages in order to ensure a strong flow of labor from the predominantly rural population. In addition, the prospect of good wages served as a powerful magnet for immigration. Wages were also kept high by the availability of land under the Free Homestead Act, whereby workers who were not satisfied with their pay were tempted to strike out for the western territories. Manufacturers were obliged to strive constantly to improve the quality and capability of their machines in order to maintain a profitable balance between the human and mechanical costs of a product. Furthermore, as machines were improved, the quantity of labor required for manufacture was reduced to the same degree that its quality was increased, enabling manufacturers to increase production at the same time as it justified the payment of higher wages.

The workers who elected to remain in the factories also played an important part in the refinement of the machines and processes for which they were responsible. When a worker was assigned to one task in the sequence of manufacture, the act of repetition increased his speed and accuracy, especially if his income was based on quantity as well as quality. The inducement of higher wages encouraged workers to be alert to any refinement in machine or process that would increase production. Alexander Hamilton had been aware of the new approach to manufacturing when he acknowledged the divisions of labor and the extended use of machines as two of the circumstances by which manufacturing could augment the revenue of society.

Many decades would pass before the number of laborers available would so exceed demand that workers would look upon the machine as their competitor rather than as their collaborator. Nor were workers concerned in the early days of manufacture, as were some social philosophers, with the danger of depersonalization. They were not concerned about the fact that they, like the machines they worked with, remained fixed while the product went by, and they were not discomforted by the prospect of being worn out or obsoleted. It was more in keeping with the times to consider work in the factory as a path to success that rewarded the laborer for his allegiance. Alexis de Tocqueville believed that Americans worked harder because they realized that success lay in their own hands. Factory workers were emerging as a new middle class that was not only able to purchase the products of industry but also enlightened enough to demand additional domestic comforts and labor-saving devices. The depersonalization of both workers and products seemed, at least at the time, a small price to pay for the better life in America.

Equality and Opportunity

The artisan . . . in an aristocracy would seek
to sell his workmanship at a high price to the
few; he now conceives that the more expeditious
way of getting rich is to sell them at a low price
to all.

Alexis de Tocqueville, 1835 (86, II, 58)

One of the tenets of the American democracy
was the conviction that everyone had an equal
right to strive for a better life for himself and
his dependents. By 1829, under the Jacksonian
banner "Let the People Rule," Americans felt free
to grasp every opportunity that would provide the
greatest reward with the least consumption of
time and energy. Klemm has written that "the
agents of this introduction of technology were the
middle class," whose "minds were filled with . . .
liberalism." (54, 269)

While the working classes in Europe were
destroying machines because they were "not apt
to distinguish between present inconvenience
and permanent evil" (192, 232), the Americans
welcomed the industrial revolution as an essential complement to the democratic way of life.
They believed in progress in science and technology as the source of national security and
economic and social well-being, and shared an
almost naive faith in the promise of machines and
mechanized production as the most effective way
to serve their needs and desires.

James Fenimore Cooper wrote: "The question
of manufactures is clearly one of interest. Of their
usefulness, and of their being one of the most
active agents of wealth, as well as of the comfort
of society, there can be no doubt. . . . Fifty years
ago, they manufactured next to nothing. They
now manufacture almost every article of familiar
use, and very many of them, much better than
the articles that are imported." He described the
population as having a "respectable degree of
intelligence" whereby the necessities of life provided an incentive for invention and talent. "Their
wants," wrote Cooper, "feed their desires, and
together they give birth to all the thousand
luxuries of exceeding ingenuity that are wanted
elsewhere." (22, II, 320) It was evident to Cooper
and others that the American home would become the focus for the good life, and therefore

that new methods should be devised to provide
individual homes for more people, that labor
within the home should be made more endurable,
and that the arts of living would become at least
as important as the struggle for existence.
Cooper concluded: "Manufacturing is a pursuit
so natural, and one so evidently necessary to all
extended communities, that its adoption is inevitable at some day or other. . . . If it be admitted
that a people, who possess the raw materials
in abundance, who enjoy the fruits of the earth
to an excess that renders their cultivation little
profitable, must have recourse to their ingenuity,
and to their industry, to find new employments
and different sources of wealth, then the Americans must become manufacturers." (22, II, 329)

The new middle class in the United States looked
to technology as the key to its equality. Industry
became the medium of their liberation, and the
acquisition of its products the symbol of their
success. In 1847 an editor of *Scientific American*
at the time applauded "this democracy which
invites every man to enhance his own comfort
and status," saying that "to the citizens of the
United States inventions are the vehicle for the
pursuit of happiness."

A Scandinavian traveling in the United States
reported that schoolboys amused themselves
by drawing steam engines or steamboats on their
slates. Today it would be said that American
schoolboys doodle rocketships on the margins
of their notebooks. Lewis Mumford once noted
dourly that if one was in love with a machine
there must be something wrong with his love life.
Yet he must have known what the machine has
meant to Americans and understood that the love
expressed is not that of human passion or com-
passion but rather a respectful admiration for the
inanimate forces of nature as they have been
transformed to human servitude.

The promise that technology could be applied
to emancipate man from back-breaking and soul-
bending work, that it could make one person as
effective as fifty and yet provide work for fifty
more, supported a social revolution that was
consummated in the United States. By it a great

middle class emerged, secure and comfortable in its homes, with amenities that even the most noble aristocrat could not have imagined in the past. Some social philosophers of the period, however, saw in the machine a force for the depersonalization of humans. Thomas Carlyle reflected Schiller's fear that the machine had reduced man to a fragment of the whole: "Were we required to characterize this age of ours by any single epithet, we should be tempted to call it not an Heroic, Devotional, Philosophical, or Moral Age, but above all others, the Mechanical Age. It is the Age of Machinery, in every outward and inward sense of the word; the age which, with its whole undivided weight forward, teaches and practices the great art of adapting means to ends." (15, I, 465) (A hundred years later the second decade of the twentieth century would be called the Machine Age, but the second time around the epithet would be used in awe rather than in condemnation and would serve as an aesthetic stimulus for Art Moderne.)

Carlyle voiced a poetic lament to mourn the changes brought on by the machine:

. . . on every hand, the living artisan is driven from his workshop to make room for a speedier, inanimate one. The shuttle drops from the fingers of the weaver and falls into iron fingers that ply it faster. The sailor furls his sail and lays down his oar and bids a strong, unwearied servant, on vaporous wings, bear him through the waters. . . . There is no end to machinery. Even the horse is stripped of his harness and finds a fleet fire-horse yoked in his stead. . . . For all earthly, and for some unearthly purposes, we have machines and mechanic furtherances. . . . We remove mountains, and make seas our smooth highway; nothing can resist us. We war with rude Nature, and by our resistless engines, come off always victorious, and loaded with spoils. (15, I, 465)

Instead of serving the cause of equality, Carlyle warned, machines would increase the distance between the rich and the poor because the Age of Machinery would create an industrialized society regulated by cold, unfeeling social systems. "Men are grown mechanical in head and in heart as well as in hand," he wrote. "They have lost faith in individual endeavor, and in natural force, of any kind." (15, I, 468) Carlyle feared

that fascination with machines had led men to an unnatural morality that considered pleasure and profit virtues, and he proposed that the inward human force that embraced spiritual and emotional values was being threatened by the outward force that was preoccupied with mechanical principles. Carlyle supported the traditional point of philosophy that the machine ethic must be brought into balance with human values.

However, the North American continent offered a special opportunity for the development of technology that was not present in Carlyle's England and in other older countries that were as overpopulated as they were short of natural resources. Once the Americans had made the ideological shift away from Jefferson's dream of a pastoral paradise to the concept of nature as a source of energy and raw materials, industrial growth was inevitable. Their acceptance of this transformation coupled with their faith in the ability to control their own future established the unique American concept of technological progress.

The American responses to Carlyle came quickly in a series of articles in the *North American Review,* a magazine that supported the rationalists and transcendentalists of New England. The first came from a young attorney, Timothy Walker, who in an article entitled "Defence of Mechanical Philosophy" set aside Carlyle's fears that the "mind will become subject to the Laws of Matter, that physical science will be built upon the ruins of our spiritual nature, that in our rage for machines, we shall ourselves become machines." Instead, Walker proposed that "the more we can compel inert matter to do for us the better it will be for our minds, because the more time we shall have to attend to them." He concluded that Americans looked "with unmixed delight at the triumphal march of mechanism," which, far from enslaving, had "emancipated the mind, in the most glorious sense." (201, 123) Walker promised that that nation that devised the greatest number of labor-saving machines would make the greatest intellectual progress.

Later articles in the *North American Review* expanded Walker's thesis that machinery had increased the amount of useful energy by increasing the products of industry. An even more important observation was that, with mechanization, man's "intellectual condition is greater" because "knowledge is diffused widely through all classes of society." The democratization of information was recognized as a great force in the development of a middle-class society in America. As a result, knowledge lost its preciousness but not its value, and a "universal public opinion" (201, 239) was formed that "has strength in its own nature. . . . It dethrones kings, it abrogates laws, it changes custom." (201, 242)

Thus, scarcely 20 years after the phrase "industrial revolution" was coined in France, the debate between human and machine values had been joined. Ralph Waldo Emerson managed to bracket the question. His ode to Channing sets the phrase "Things are in the saddle, / and ride mankind" (63, 178) against an acceptance of that "marvelous machinery which differences this age from any other age." (63, 263)

The American hunger for utilitarian products stimulated the emergence of a class of entrepreneurs, designers, and inventors—the agents of technological progress. Every American hoped to become a "budding capitalist" as his initiative and imagination conceived products and product systems that were suited to increased manufacturing capability. And as production volumes went up and the cost per manufactured unit went down, industries began to direct their attention to creating the marketing institutions and distribution systems that were needed to absorb the flow of products from their factories. It is more than a coincidence that communication and transportation evolved side by side with industrialization.

It was in this atmosphere of progress that the turnpikes, canals, and railroad systems of the United States were established. From a very slow start in the colonial period, carriage building developed into a major industry producing overland stages, city omnibuses, cabs and hacks, Conestoga wagons, prairie schooners, freighters, elegant private buggies, runabouts, trotting wagons, phaetons, sulkies, and buckboards. Carriages were steadily refined to reduce material and weight in order to increase relative horsepower, and an undeniable elegance was the result.

Canal technology was perfectly suited to the early nineteenth century. Immigrants were employed to dig the ditches, to quarry the limestone on the hillsides along the projected waterway, to construct the kilns in which to burn the limestone into mortar, to sled the stone down the hills to line the ditch, and then to build the boats, locks, and bridges with timber from the surrounding forests. When the most successful American canal, the Erie, was inaugurated in 1825 with a voyage from Buffalo on Lake Erie across New York State to the Hudson River and thence down the Hudson to New York City, it reduced the time of transportation between the Great Lakes and the Atlantic Ocean by half. The cost of freight transport was so low—one-fifth that of transport by wagon—that the Erie remained in service for 100 years until it was replaced by the New York Barge Canal. Though the original canals have passed into history, they have left their mark on the landscape in the form of stonework and artifacts that exhibit the reserved functional aesthetic of the era.

The steam locomotive and railroads that followed the canals were somewhat higher on the technological ladder; whereas canal equipment was built of wood and stone with iron as a tectonic material, the railroads put iron first with wood and stone coal secondary. The first steam engines built for land locomotion were used primarily in the coal-mining industry to replace horse-drawn wagons on wood or metal rails. No practical passenger rail transportation was developed until the second quarter of the nineteenth century, by which time the steam engine had been reduced in size and made substantially safer. In 1829 the Manchester Railway, employing the locomotive *Rocket* built by George Stephenson, opened the first passenger steam railway in England, connecting the port of Liverpool with the cotton-manufacturing city of Manchester. Even though the United States had passed its first railway act in 1823, incorporating the Pennsylvania Railroad Company, the first locomotives to operate there were English: the *Stourbridge Lion* imported in 1829 and the *John Bull* in 1831. The first American locomotives were the *Phoenix* and the *West Point,* built in 1831 by the West Point Foundry. Once started, how-

Concord coaches were built from 1825 (when Louis Downing and J. Stephen Abbot began manufacture in the New Hampshire town that gave them their name) until they were displaced by steam trains and electric trolleys late in the century. Despite conventional associations with Western movies, they served best as the primary means of public conveyance on the post roads of the East. Harper's Weekly, *August 1870.*

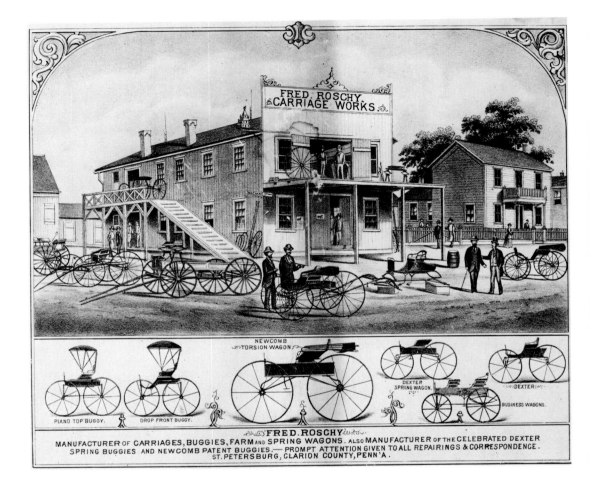

Fred Roschy's carriage
works at Petersburg in
Clarion County, Pennsyl-
vania, was equipped to
produce a broad variety
of the elegantly spare
wagons and buggies of
the era—the counterparts
in transportation of Shaker
furniture. Library of
Congress.

In the early years of the
nation, before overland
turnpikes were joined into
dependable roadways,
rivers and canals provided
the basic transportation
arteries for people and
their raw materials and finished products. The Ka-
nawha Canal, which par-
alleled the James River in
Virginia, was one of the
most important. Harper's
Weekly.

George Stephenson's
Rocket *locomotive initi-*
ated railroad transpor-
tation by winning a
competition for the best
steam locomotive to run
on the new Manchester
(England) railroad. On the
same day, the Rocket *ran*
over a bystander, causing
the first railroad fatality.
Byrn, The Progress of
Invention in the Nineteenth
Century.

The "West Point" locomotive and its train of cars was among the first to be built and operated in the United States. This drawing, although ceremonial in character, illustrates the stark simplicity and dry elegance of the purely functional technological product. Smithsonian Institution.

ever, railroad systems spread rapidly across the country. By 1832, three years after the first locomotive was put into service, over 1,400 miles of railway had been laid and were in use, signaling the eventual eclipse of the canal systems in the United States.

M. W. Baldwin, one of the most successful American locomotive builders, started out as a jeweler who was commissioned to build a miniature steam locomotive for Peale's Museum in Philadelphia. It is interesting to speculate that full-size locomotives may often have been no more than enlarged versions of the original models. This egg-before-the-chicken routine was encouraged by the fact that for many years the United States Patent Office required a model before allowing a patent.

As the first great means of land transportation based on technology, the locomotive became the principal symbol of the age much as the airplane was to become the symbol of the twentieth century. To some philosophers of the time, locomotives were man-made monsters, belching smoke and cinders to befoul the environment. Thoreau feared the railroad: "We do not ride upon the railroad, the railroad rides upon us." To others, locomotives were dramatic symbols of man's mastery over nature, space, and time. With greater distances to span than its European counterparts, the American locomotive became leaner, longer, and more powerful, with graceful form and a disposition of detail that "gave a finished appearance to the whole, sufficient to raise it to the dignity of a work of genuine art." (11, III, 322)

American locomotives were often decorated with oil paintings, gilt and brass trimmings, and cast-iron figures. This practice, which reached its zenith in the eloquent extravagance of the great steam-driven paddle-wheelers that were beginning to ply the inland American waterways, was to filter down to all labor-saving machines. Although the contribution of the artist-designer to the locomotive and other technology-based products added nothing to speed, dependability, or efficiency, it was indispensable to the pride of the builders and the expectation of the con-

sumers. This practice of ornamentation is in no way different from the use of colorful "streamers" and extravagant graphic symbolism to decorate contemporary jet airliners. Nevertheless, beneath the costume of communication, there still remains the essential form of the product upon which the final communication between a man and his product depends. When the paint has worn off and when the trim and the decorative details have weathered away, the substantive meaning of the product is revealed. It is no different for the objects of technology than it is for other products of man.

There must have been hundreds of designers involved in the process of conceiving and refining the many utilitarian products that preoccupied early industry, but they were and remain essentially anonymous. In the young and virile democracy it was thought that everyone was, or may have been, a contributor to the evolution of a product. Only on rare occasions was a concept so historic and distinctive that its creator was entitled to public honor. And even then, as the painful trials of men like Eli Whitney, Samuel F. B. Morse, Elias Howe, and Cyrus McCormick illustrate, the rewards were tempered with anguish. It would seem that anonymity was then, and continues to be, part of the burden that designers must bear for their role in democratic utilitarian design. The products themselves were often more honored than their creators. They were held up to the public and unashamedly identified by titles that enhanced their images. Locomotives bragged about their power with names like *Sampson* and *Champion,* their speed with *Flyer* and *Mercury,* or their patriotism with *Washington* and *America,* and the great American clipper ships sped to glory with names like *Hurricane* and *Lightning.*

Of all of the transportation products of the mid-1800s, the clipper ship seems to have the greatest hold on the American imagination. (Witness the impact of the Tall Ships' visit to New York harbor on July 4, 1976.) For 21 years in the middle of the nineteenth century these "canvas-backs," in a final flash of eighteenth-century technology, were able to outrun every other vessel on the high seas—steam or wind driven.

The first locomotive Baldwin built was Old Ironsides, *constructed in 1832 at his Philadelphia works. It provided service for more than 20 years at a top speed of 30 mph drawing a full train. Byrn,* The Progress of Invention in the Nineteenth Century.

Baldwin's Tiger *locomotive, offered for sale in 1856. URS – Coverdale & Colpitts.*

Paddlewheelers along a New Orleans levee in 1860. Library of Congress, Eskew Collection.

Yet the designer of the clipper, John W. Griffiths, expressed concern over the fact that his first clipper, *Rainbow,* had been called beautiful. "We do not understand the import of the term 'beauty,'" he observed in his *Treatise on Marine and Naval Architecture.* "We can give no other definition than the following: fitness for the purpose and proportion to effect the object designed." Griffiths had extended the eighteenth-century French experiments in hydrodynamics into a theory that the traditional form of a ship's hull was in error. Until Griffiths, it had been thought that the widest part of the hull should be well forward, the bow well rounded, and the stern narrow. This design had been based on the erroneous idea of designing ships after the lines of fish. Griffiths's hull for the clipper turned the fishes' proportions around, making the bow the narrowest part (with a negative curve and an overhanging deck) and moving the widest part of the beam aft of center. This philosophy of form in the interest of speed was lost on the early promoters of streamlined trains and automobiles in the early 1930s, who held up the teardrop or comet shape as the most aerodynamic form.

Griffiths had demonstrated once again that the ideal form of a product must be that which serves its purpose most efficiently, but could not quite accept the principle of design that finds beauty as the natural result of a clear solution to a problem. The leading proponent of this concept of beauty through function and of humility before the laws of nature was Horatio Greenough (1805–1852), a natural philosopher and essayist who, although he worked as a sculptor, would be remembered more for his aesthetic perception than the quality of his statues. He was a virtual expatriate who sensed from afar the cultural spirit in the American struggle for survival and dignified the insoluble bond between beauty and utility. Greenough was the spiritual example for Louis Sullivan, Frank Lloyd Wright, and the entire school of organic architects and functionalist designers who elected to solve problems by the most direct means and let beauty come when it may. In his essay on *Aesthetics in Washington* he complimented "the men who have reduced locomotion to its simplest elements, in the trotting wagon and the yacht, *America*" (the racing

Donald McKay and John Griffiths were the most renowned designers of clipper ships. McKay's Flying Cloud, *built in East Boston in 1851, went on to set records for speed and performance under weather and was considered by many to be the most beautiful clipper. Museum of the City of New York, Harry T. Peters Collection.*

The figurehead for the ship Edinburgh, *carved in oak by John Rogerson, is representative of the attention paid by shipbuilders to a decorative detail that helped give a vessel her identity. Index of American Design, National Gallery of Art.*

Horatio Greenough's
heroic portrait of George
Washington as a noble
Roman was created on
an assignment from Con-
gress to grace the Capitol
in Washington. Its classical gesture and costume,
as shown in this con-
temporary lithograph,
were too much, even for
Congress. Author's
collection.

schooner that defeated the vessels of the Royal
Yacht Squadron in a race off the Isle of Wight in
1851) as being "nearer to Athens at this moment
than they who would bend the Greek temple to
every use." Greenough contended for "Greek
principles, not Greek things." "If a flat sail goes
nearest wind," he wrote, "a bellying sail, though
picturesque, must be given up." On carriage
design, he wrote that "the slender harness and
tall gaunt wheels are not only effective, they are
beautiful—for they respect the beauty of a horse
and do not uselessly task him." (40, 22) Greenough was pleased with his proof of the perfect
adaptation of ships to their function. In his essay
Travels of a Yankee Stonecutter he observed
of figureheads on clipper ships "that the only part
of the hull where function will allow a statue to
stand without being in Jack's way, is one where
the plunge bath so soon demolishes it." (41, 179)

Horatio Greenough set the foundation stones
for the American design ethic with the eloquent
proclamation "Beauty as the Promise of Function;
Action as the Presence of Function; Character
as the Record of Function." (40, 71) And, while
apologizing for separating elements of a common
principle, he took each as a phase through which
organized intention passes to completeness.
He contended that in absolute dedication to
function was to be found the quintessence of
beauty. Though Greenough himself had been
caught up in the style of the moment when on
order from Congress he had produced a heroic
marble statue of George Washington half-clad
in a Roman toga for the nation's capital, he
deplored the senseless application of style to
architecture and called for an organic architecture
whose external form would be determined by
internal need. His primary concern was with the
propriety of classical form applied to democratic
architecture, and he was probably unaware of the
dual revolution in the technology rather than the
aesthetics of building that was underway by
mid-century.

The first aspect of this revolution was a structural
system conceived by carpenter-builders in the
1830s that depended upon power-sawn standardized lumber and standard nails produced by

newly invented machinery. It reduced the art of building from cabinetry to carpentry by replacing pegged mortise-and-tenon joints with overlapping connections and diagonal bracing. The "balloon" system of building, as it was derisively referred to at the time, was originally intended as a convenient and cheap way to put up temporary shelters. However, balloon-framed buildings proved to be so strong and durable that the system displaced traditional methods. It was well-suited to the America of its time, with forests available for the cutting, powered machinery for cutting lumber and making nails, and low-cost immigrant labor that only needed to know how to saw a board and hammer a nail. Many new books showed the variety of buildings that were possible with the system, including Edward Shaw's *The Modern Architect* (1855) and Lewis F. Allen's book of designs for balloon-frame houses, *Rural Architecture* (1852). Allen supported his philosophy of structure by stating that good taste demands "a fitness to the purpose for which a thing is intended, . . . a harmony between its various parts," and that any product of good taste would be "pleasing to the eye, as addressed to the sense, and satisfactory to the mind, as appropriate to the object for which it is required." (2, 48) The wooden houses of the United States combine eighteenth-century forms with nineteenth-century techology. They attracted worldwide attention in the 1860s and were even shipped in sections to be exhibited in European expositions as a curious complement to American native ingenuity. (It should be noted that, like most modular systems that promise infinite variation, balloon-framing has fallen victim to its own framework. It has resulted in a Gordian knot of specialized occupations, each coated with layers of defensive industry and government controls. Thus, any system that may have been conceived originally as an inspired methodology along the cutting edge of technology tends to become less flexible in time and slowly slips back to become a deterrent to progress in the very area that it served so brilliantly in the beginning.)

A second new building system came into being in 1842 when Daniel D. Badger erected a storefront in Boston from cast-iron sections. Cast iron had become a popular material for stoves, fountains, and garden furniture, and its architectural application opened the way for "a style of street architecture, as applied to retail shops, of a different character to that which now prevails, and which is in imitation of European modes alike of construction and decoration." (76, 264) James Bogardus of New York must be credited with building, in 1848, the first full building facade of cast iron using manufactured interchangeable parts. Again, the system was admirably suited to the technology of the times since it depended largely on low-cost, primarily immigrant labor to carve the wooden patterns, cast the molten iron, put the parts together, and then paint the whole to look as though it had been built of wood and stone. In 1856 Bogardus described its advantages: "Such a building may be erected with extraordinary facility and at all seasons of the year. No plumb is needed, no square, no level. As fast as the pieces can be handled they may be adjusted and secured by the most ignorant workmen. The building cannot fail to be perpendicular and firm. . . . It follows that, a building once erected, it may be taken to pieces with the same facility and dispatch." (12, 7)

Within a decade a controversy was raging among architects over the threat of manufactured buildings. A court case had just been won in which architecture had been acknowledged as a profession that commanded a fee, and members of the newly founded American Institute of Architects (1856) were understandably sensitive about their prerogatives in directing the course of architecture in America. On one side of the argument were those who saw any mechanized and industrialized method of building as a threat to the traditional arts of designing unique structures of stone and wood—particularly since the finished forms of the structures were in no obvious way different from the historic styles of architecture upon which they themselves depended for inspiration. On the other side were those who wanted the architects to participate in the new technology. Henry Van Brunt read a paper to the AIA that put forth this conviction: "This is called an Iron Age . . . for no other material is so omnipresent in all the arts of utility. . . . But architecture, sitting haughtily upon her

The balloon frame, so called because of the lightness of the structure, was apparently conceived around 1835 by George W. Snow of Chicago as a method of prefabrication to take advantage of standardized lumber sizes, manufactured nails, and unskilled labor. After a century and a half it continues to dominate the building of private houses in the United States, despite repeated promises of mass-produced homes. Metropolitan Museum of Art.

The first full-sized building constructed from cast-iron standardized parts by James Bogardus was erected in 1848 at the corner of Centre and Duane Streets in New York. Museum of the City of New York.

Acropolis, has indignantly refused to receive it, or, receiving it, has done so stealthily and unworthily, enslaving it to basest uses, and denying honor and grace to its toil." (21, 79)

Despite their innovative technology, these two new systems depended in part for acceptance upon prevailing architectural style. Balloon framing was (and still is) the preferred method for domestic architecture built primarily in the colonial style. And, although the cast-iron system evolved into the welded steel skeleton, commercial buildings continued for decades to be clothed in historic style.

The latter half of the nineteenth century was aptly called the Iron Age by Van Brunt. In its cast and wrought forms, iron was the most challenging building material of the time. Its potential was dynamically demonstrated by the main structure of the great International Exposition held in London in 1851.

The concept of industrial exhibitions was invented by the French in the aftermath of their revolution. When it was found that workmen were starving despite the fact that storehouses were filled with wares, the Marquis d'Aveze obtained permission to hold a public exhibition in the garden of the Maison d'Orsay in an attempt to sell the overstock. The success of the sale led to the first official exposition, held in 1798 in the Champs de Mars in a "Temple of Industry" erected for the purpose. The exhibition attracted particular attention by inaugurating a system of jury awards for excellence in design and workmanship. After that, national exhibitions were held for half a century, not only in France but also in other countries. Then, in 1849, when the French abandoned a recommendation to expand their fair to international status, the concept was picked up by the British.

In 1849 Prince Albert of England invited other countries to participate in an exhibition in which prizes would be awarded to outstanding work in four great divisions: raw materials, machinery and mechanical inventions, manufactures, and the plastic arts. After the original competition for an appropriate building failed to produce a struc-

ture deemed worthy of the scale of the exposition, it was decided to accept a daring new concept proposed by Joseph Paxton that was based on his experience with iron structures and glass in building greenhouses. The Crystal Palace was, in effect, an inspired modern adaptation of the traditional cruciform with its naves and transepts—"designed, however, in a new style of architecture; not massive, dark and somber, but light, graceful, airy and almost fairy-like in its proportions. . . . a true 'Crystal Palace,' and a noble example of the use of our modern material—iron—for building purposes." The spectacular success of the exhibition, financially and otherwise, established the prototype for a series of similar international exhibitions that lasted well into the present century. The first several, including the first American Exhibition in 1853, featured unabashed copies of the Crystal Palace.

Although the London exhibition was meant to demonstrate the superior capability of the British, some critics warned that it had helped the foreigners, especially the Americans, more than the British.

The American products that were exhibited were criticized at first as being severe and even tasteless, with little or no ornamental value. The official catalog of the exhibition contained the following statements: "The expenditure of months or years of labour upon a single article, not to increase its intrinsic value but solely to augment its cost or its estimation as an object of *virtu,* is not common in the United States. On the contrary, both manual and mechanical labour are applied with direct reference to increasing the number or the quantity of articles suited to the wants of a whole people, and adapted to promote the enjoyment of that moderate competency which prevails among them." (76, 17) This was, at the very least, grudging admission that American manufactured products that were contrived to serve some specific human need were achieving respectability and a character that was distinct from those of other countries. American pianos, epergnes, carriages, and other aspirations to culture were passed over for ice-making machines, cornhusk mattresses, fireproof

The great cast- and
wrought-iron Crystal
Palace in London was a
dramatic prophecy of ar-
chitecture of the future.
The Palace was manu-
factured on the spot by
a system of traveling
scaffolding, with each
element thoroughly tested
to ensure its safety. Ewing
Galloway.

An American carriage
by Clapp and Son of
Boston that was exhibited
at the Crystal Palace. The
English catalog cited this
as a demonstration that
the Americans were not
"insensible to the luxuries
and conveniences of life."
The Crystal Palace Exhi-
bition Illustrated Catalogue.

A pianoforte designed and manufactured by Nunn and Clark of New York and shown at the Crystal Palace. Noted the catalog: ". . . as [the Americans'] wealth increases so also does their taste for the elegant and the beautiful and their desire to possess what will minister to the refinements of life." The Crystal Palace Exhibition Illustrated Catalogue.

The American Chair Company of New York exhibited this novel rotating chair, whose seat was supported by flexible springs. The Crystal Palace Exhibition Illustrated Catalogue.

Charles Goodyear's exhi-
bition of hard India rubber
goods at the Crystal Pal-
ace. Goodyear was caught
up by the "India-rubber
fever" in 1835, and determined to invent a process
that would cure the ma-
terial's inability to with-
stand either cold or hot
weather. He invented vul-
canization by accident,
launching an industry that
in short order employed
60,000 and made over 500
different products. Boston
Public Library.

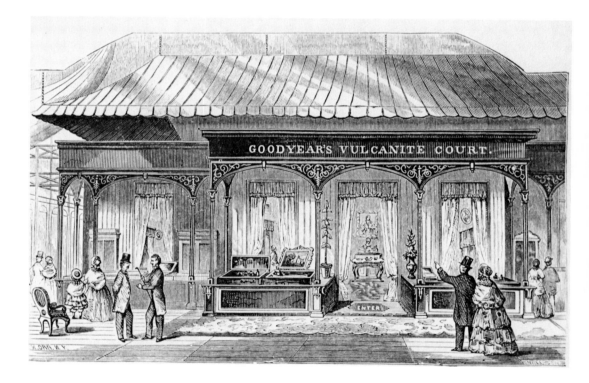

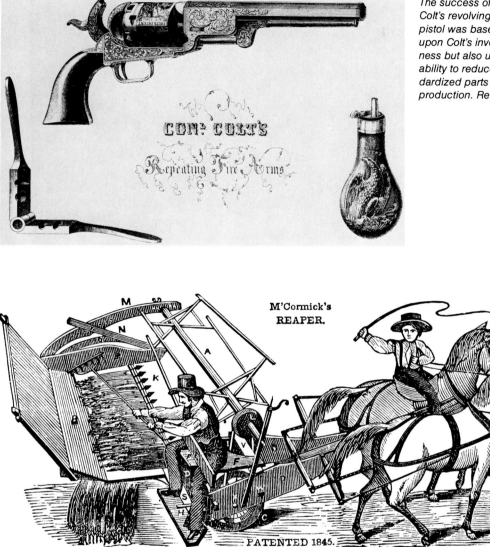

The success of Samuel Colt's revolving-barrel pistol was based not only upon Colt's inventiveness but also upon his ability to reduce it to standardized parts for mass production. Reference 75.

M'Cormick's REAPER.

PATENTED 1845.

M'CORMICK'S REAPER.

Cyrus McCormick caught the fever of invention from his father and determined to develop a reaper superior to all others. By the time he was 22 he had demonstrated a machine that was six times more effective than cutting grain with a hand scythe. His invention won a grand prize in London and made him a millionaire before he was 40. State Historical Society of Wisconsin, McCormick Collection.

safes, meat biscuits, lifeboats, railroad switches, nautical instruments, artificial eyes and legs, and such technological wonders as Charles Goodyear's vulcanized rubber products, Samuel Colt's revolver, and Cyrus McCormick's reaper.
The London *Times,* which had derided the reaper as "a cross between a flying machine, a wheelbarrow and an Astley chariot," later acclaimed it more than any other product displayed: "The reaper will amply remunerate England for her outlay connected with the Great Exhibition."

The rudest shock to British pride, however, was the victory of the yacht *America,* which "appeared among the English yachtmen like a phantom; so different in build and appointments from their own craft, that they at once came to the conclusion that if 'she was right, they were all wrong' " (75, 10), over the Royal Yacht Squadron at Cowes on August 23, 1851. Since that day, the America's Cup has been displayed proudly in the New York Yacht Club, as American yachts have emerged victorious in every contest.

The London *Times,* characterized by Daniel Webster as "the bitterest, the ablest and the most Anti-American press in all Europe," summarized the surprising achievements of the Americans a few weeks before the close of the exhibition:

It is beyond all denial that every practical success of the season belongs to the Americans. Their consignments showed poorly at first, but came out well on trial. Their reaping machine has carried conviction to the heart of the British agriculturalist. Their revolvers threaten to revolutionize military tactics as completely as the original discovery of gunpowder. Their yacht takes to a class to itself. Of all of the victories ever won, none has been so transcendant as that of the New York schooner. Besides this, the *Baltic,* one of the Collin's line of steamers, has "made the fastest passage yet known across the Atlantic".... So we think, on the whole, that we may afford to shake hands and exchange congratulations, after which we must learn as much from each other as we can. (150, 4)

The *London Observer* summarized the manufacturing philosophy of the Americans as follows:

They produce for the masses, and for a wholesale consumption. There is hardly any thing shown by them which is not easily within the

The schooner America, *designed and built by George Steers in 1851 for a group of members of the New York Yacht Club, raced and beat the Royal Yacht Squadron in her first trip across the Atlantic. Smithsonian Institution.*

The U.S. mail steamship Baltic *was built by the Novelty Iron Works of New York in 1850 for the Collins Line, and competed in a transatlantic race with the English Cunard Line. With government subsidies, Collins managed to break all records. However, high costs and ocean tragedies cost the Americans the race in the end. Reference 75.*

reach of the most moderate fortune. No Government of favoritism raises any manufacturers to a pre-eminence, which secures for it the patronage of the wealthy. Everything is entrusted to the ingenuity of individuals, who look for their reward in public demand alone. With an immense command of raw produce, they do not, like many other countries, skip over the wants of the many and rush to supply the luxuries of the few . . . they have turned their attention eagerly and successfully to machinery, as the first stage in their industrial progress. They seek to supply the shortcomings of their labor market, and to combine utility with cheapness. The most ordinary commodities are not beneath their notice. . . . (75, 127)

Two years after the London Exhibition closed the British sent two observers, Joseph Whitworth, a technologist famous as a builder of machine tools, and George Wallis, the headmaster of a Government School of Art and Design in Birmingham, to New York City to visit the American "Crystal Palace" exhibition in order to report on the nature of American markets. Because the opening of the exhibition had been delayed, they elected instead to tour American cities and manufacturing centers. Their observations recognized the value of the American system of manufactures and signaled a counterflow of technical knowledge from the United States to Europe. Wallis in particular was impressed by the differences between American and European consumers and their effect on manufactures. He wrote that the American consumer typically expected an article to last for only a short time and was therefore more reluctant to pay a premium for either quality or durability. The tastes of American consumers were more homogeneous than those of Europeans and were concentrated at the low end of the quality spectrum. The taste for short-lived articles, Wallis reported, "is said to run through every class of society, and has, of course, a great influence upon the character of goods generally in demand, which . . . are made more for appearance, and less for actual wear and use, than similar goods are in England." (76, 304) A taste bias in favor of homogeneous low-quality goods of short life expectancy was, of course, highly favorable to the introduction of mechanical techniques.

Whitworth and Wallis also observed that the Americans were reluctant to incur additional costs in ways that did not improve the efficiency of the final product in some narrowly defined utilitarian sense. While they commended the native ingenuity, energy, and perseverance in the development of machinery and utilitarian products, they deplored the cupidity of merchants who often force the manufacturer not only to copy foreign designs but also to mark the product with the name of the original maker. Their important report foreshadowed the development in the United States of a "common market" philosophy of products that was not to appear in Europe until the next century. Under it, manufactured products are generalized in character and adjusted in quality until they find the broadest market commensurate with public expectation, need, and ability to pay.

Elegance and the Middle Class

Our progress in these modern times then, consists in this, that we have democratized the means and appliances of a higher life–that we have brought and are bringing more and more the masses of the people up to the aristocratic standard of taste and enjoyment and so diffusing the influence of splendor and grace over all minds.

Horace Greeley, 1853 (39, 52)

In its first years the American republic was led mostly by men who, although they supported and defended the idea of equal privileges for all, had been themselves members of a colonial elite. The public posture of the Americans may have called for asceticism and self-denial, but their private dreams were to share the elegance and luxury of the personal and environmental furnishings of the higher classes of Europe. Those who could afford to do so satisfied their desires by importing original products from abroad. Those who could not had them duplicated in America, primarily by émigré English, French, and (later) German designers and craftsmen.

American manufacturers also had to "contend with one difficulty," as James Fenimore Cooper observed, "that is not known to the manufacturers of other countries." Wrote Cooper:

The unobstructed commerce of the United States admits of importations from all quarters, and of course, the consumer is accustomed to gratify his taste with the best articles. A French duke might be content to use a French knife or a French lock; but an American merchant would reject both: he knows that the English are better. On the other hand, an English dutchess (unless she could smuggle a little) might be content with an English silk; but an American lady would openly dress herself in silk manufactured at Lyons. The same is true of hundreds of other articles. The American manufacturer is therefore compelled to start into existence full grown, or nearly so, in order to command success. I think this peculiarity will have, and has had, the effect to retard the appearance of articles manufactured in this country, though it will make their final success as sure as their appearance will be sudden. When the manufacturers of America have once got fairly established, so that practice has given them skill, and capital has accumulated a little, there will be no fear of foreign competition. (22, II, 328)

However, a weakness developed in American manufactures that still persists in most of the industrial arts. The American climate of open competition fed upon the ethnic ties and cultural nostalgia of the Americans in a way that gave imported articles an advantage in the marketplace. Therefore, manufacturers in this country found it to their advantage to copy the designs of the most popular imports rather than strike out on an uncharted course of their own. They had no need of trained American designers, it appeared, so long as each boat landing brought in the latest sources of aesthetic inspiration.

This circumstance had quite the opposite effect on those countries of Europe that were becoming increasingly dependent upon exports to the United States. They realized that they would have to produce goods that were consciously designed to be more attractive to foreign buyers. The artistic studies that had been part of the mechanics institutes were pulled out and organized as separate design schools in several European countries during the second quarter of the nineteenth century, primarily for the purpose of training artists for industry. In Great Britain the principles of industrial art (defined as manufactured products in which elements of art and style are essential to marketing) were championed by Henry Cole, who set out to show that "an alliance between fine art and manufacture would promote public taste" and that "elegant forms may be made not to cost more than inelegant ones." Parliament assigned a select committee to examine "the best means of extending a knowledge of the arts and principles of design among the people," and as a result the Normal School of Design (later to be known as the Royal College of Art) was established in 1837. The following year a committee of interested citizens in the important English manufacturing center of Manchester recommended that, since mechanics were already being excluded from classes at the design school in London, Manchester too should have a separate school for designers. Though they acknowledged that the mechanics institute in Manchester offered several courses in artistic drawing and hoped that these would be continued, they recommended "the formation of a society having for its sole and peculiar object to improve the arts of design, an object sufficient to occupy the whole time and attention of a society with reference to the improvement of those manufactures in which design is required; and also in the education of persons to direct the mechanical powers of this great community." The committee further noted that "superiority in manufacturing depends, in a great measure, on the fortunate exercise of taste, economy, industry and invention," and recommended the establishment of a school of design in Manchester "in order to enhance the value of the manufactures in this district, to improve the taste of the rising generation; to infuse into the public mind a desire for symmetry of form, and elegance of design; and to educate, for the public service, a highly intelligent class of artists. . . ." (134, 129)

In 1840, President Alexander Dallas Bache of the Franklin Institute in Philadelphia described in the *Journal* of the Institute the program and structure of the Institute of Arts in Berlin, making particular note of the fact that it was supported by the combined efforts of government and private interests. He concluded his report with the expressed hope that "it will not be very long before the institution of seminaries, analogous in principle . . . will become an object of legislative regard in some, at least, of the United States. . . ." (135, 46) And, from abroad, the essays of Horatio Greenough lent support to Bache's plea for schools of design in America. He realized that the older countries of Europe had founded and supported schools of design not because they had been "invaded by a sudden love of the sublime and beautiful." "I believe," Greenough wrote, "that they who watch our markets and our remittances, will agree with me, that their object is to keep the national mints of America at work for themselves; and that the beautiful must, to some extent, be cultivated here, if we would avoid a chronic and sometimes an acute tightness of the money market. The statistics of our annual importation of wares, which owe their preference solely to design, will throw a light on this question that will command the attention of the most thrifty and parsimonious

of our legislators." Greenough expressed a "desire to see working Normal schools of structure and ornament, organized simply but effectively, and constantly occupied in designing for the manufacturers, and for all mechanics who need aesthetical guidance in their operations." He was not certain to what extent the central government would be willing to support such a school, but expressed a hope that if it did not some of the states or private individuals would step in. Greenough suggested that a study of the matter "called for by Congress, on the amount of goods imported, which owe the favor they find here, to design, would show the importance of such schools in an economical point of view," and stated the belief that such a report "would show that the schools which we refuse to support here, we support abroad, and that we are heavily taxed for them." (41, 21)

Neither the government of the United States nor the manufacturers of the time gave sufficient weight to the pleas of Bache and Greenough for the establishment of European-type design schools. In his 1854 report on American manufactures the Englishman George Wallis noted that only one design-related school in the United States, the Maryland Institute in Baltimore, was in any way close to the standards and objectives of European schools. The only other schools were primarily devoted to teaching young women in art as applied to the textile and wallpaper industries. The first of these was a private school, The Philadelphia School of Design, that was founded by the Franklin Institute in 1850 with courses teaching design and drawing in classical, mechanical, and oriental styles and ornament. Its success encouraged the establishment of similar schools in Boston in 1851 and New York in 1852. In general, manufacturers were aware of the value of teaching female designers to be decorators; thus they provided some support to these specialized schools and offered employment to qualified graduates.

Wallis made an interesting observation on a developing difference between European and American art education: In England, Wallis wrote, such education was directed at creating a class of artisans for industry, but the United States had

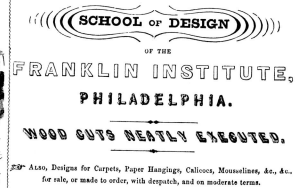

SCHOOL OF DESIGN

OF THE

FRANKLIN INSTITUTE,

PHILADELPHIA.

WOOD CUTS NEATLY EXECUTED.

☞ ALSO, Designs for Carpets, Paper Hangings, Calicoes, Mousselines, &c., &c., for sale, or made to order, with despatch, and on moderate terms.

M

To THE SCHOOL OF DESIGN, Dr.

BILLHEAD OF THE SCHOOL OF DESIGN.

From a wood cut made by a pupil in 1851.

One of the earliest schools of design, that of the Franklin Institute, was organized primarily to train young women who were deemed to be superior in skill and more sensitive to current ornamental styles. Wright, The Story of the Franklin Institute.

"reached that point at which it has become desirable that originality of thought should be infused into [manufactured products] by means of the designer." (76, 297) Nevertheless, many decades were to pass before pressure for indigenous product aesthetics was to become great enough to awaken academic institutions and manufacturing industries to support the establishment of schools of industrial design. Though a succession of studies had recommended such schools over the years, it was evident that, for the time being, the growing nation preferred to meet its needs with imported aesthetics. American talent, it would seem, was not to be judged respectable until it had been bathed and perfumed in the latest cultural fragrance of the continent. Few were able to sense that an art of democratic manufactures was taking form in America— born out of necessity and economy, with a growing affection for the spare aesthetics of utilitarian products. For many years, the unique beauty of American products was to be ignored by those to whom aesthetic value was simply a layer of style.

During the early decades of the democracy, American furnishings were primarily based on the neoclassical fashions of England and France. Furniture for the wealthy was generally imported, or handmade on this side of the Atlantic by immigrants like Charles Lannuier and Gabriel Quervelle from France and Joseph B. Barry from England, or else copied by American craftsmen from the imported design books of the Adams Brothers, George Hepplewhite, or Thomas Sheraton. Later the design books of George Smith and the Nicholson Brothers served as sources for the bolder forms and heavier style known as Greek Revival. By 1840 American cabinetmakers were producing their own style books, such as The Cabinet Maker's Assistant by John Hall of Baltimore.

The emerging middle class in America struggled first for security and convenience; however, these goals were, for the most part, only stages to a higher standard of living. The common man of the democracy was not content to remain common. Having acquired the necessities of living, he

hungered for its felicities. If he could not afford the unique treasures of the wealthy, he welcomed whatever semblance of the finer things of life that manufacturers could provide. Even so, he felt more comfortable if such domestic accessories as he purchased were identified for him as "practical luxuries." Alexis de Tocqueville captured the dichotomy of middle-class taste:

It would be to waste the time of my readers and my own, if I strove to demonstrate how the general mediocrity of fortunes, the absence of superfluous wealth, the universal desire for comfort, and the constant efforts by which everyone attempts to procure it, make the taste for the useful predominate over the love of the beautiful in the heart of man. Democratic nations, amongst which all these things exist, will therefore cultivate the arts which serve to render life easy, in preference to those whose object is to adorn it. They will habitually prefer the useful to the beautiful, and they will require that the beautiful should be useful. (86, II, 56)

Tocqueville's perceptive insight into the aspirations of the middle classes lies close to the dynamism that moves democratic design. Over the two centuries that followed the Revolution, American philosophers and designers have made repeated efforts to achieve recognition of the alliance between the useful and the beautiful. Industrial design in the United States came into being as a result of just such a demand for sensitive individuals who could provide utilitarian objects with forms that would be meaningful as well as handsome.

American manufacturers reacted to the increasing demands of the middle class by developing machinery and processes to produce larger volumes of glasswares, furniture, oilcloth and floor coverings, wallpaper, lighting fixtures, clocks, and many other domestic products. Methods of manufacture that had been suitable for low-volume, labor-intensive production were modified to meet the opposite objectives of high-volume, low-labor production.

The invention and manufacture of "pressed glass" offers a good example of American ingenuity in increasing volume by transforming the original practice of blowing glasswares and then facet-cutting them to bring out the brilliance of the glass into a process of using cast-iron molds for pressing glasswares, pattern and all. This brought the general appearance (if not the ultimate quality) of cut glass within the financial reach of the average American, and by 1834 it had stimulated a reverse flow of glass products to England. Despite complaints that such products were cheap and would degenerate public taste for decades to come, American pressed glass has been elevated to the status of important antiques.

Of all furniture, the chair quickly became the dominant symbol of middle-class gentility. As now, it served notice of the level of taste of the householder. Therefore, chairmakers were especially responsive to increasing demands. In the transformation to quantity production, hand-carved legs gave way to stiffer ones with forms that could be produced by profile-cutting lathes that ensured uniformity at the hands of relatively unskilled factory workers. Seat and back elements became flatter, generally more rectangular in shape, and decorated with stenciled painting rather than carving.

Painted furniture was not new to the democracy, but now paint was used either to simulate the more expensive woods, to conceal the mismatched grain of inferior wood, or to provide a smoother base for the stenciled decoration used on "fancy" chairs. Fancy chairs had been developed originally in imitation of oriental lacquering. However, the neoclassic style of Adams followed by Hepplewhite and Sheraton had succeeded the Chinese influence. By the time American fancy-chair manufacturers began production in the early nineteenth century, these English designers were setting the aesthetic pace. Fancy chairs were produced in great quantities over the first half of the century, and as their prices came down they found their way into private homes and public places and on to the river steamboats. A good indication of the number of fancy-chair manufacturers is the fact that no fewer than 200 took part in the great civic parade in New York celebrating the opening of the Erie Canal in 1825. The most famous of the fancy-chair makers was Lambert Hitchcock (1795–1852), whose chairs may have been the first

to be adapted to quantity production in the United States. He established a factory in 1818 in Connecticut to manufacture chair parts for others. Between 1828 and 1848 he was manufacturing his own chairs, and the settlement that grew up around his factory became known as Hitchcockville. (It was changed to Riverton in 1862, ten years after Hitchcock's death.) The original factory was reopened in 1949 in response to the demand for Hitchcock chairs. At its peak, Hitchcock's factory was producing as many as 15,000 chairs a year, some of which were sold at wholesale for as little as 25 cents each. At one time or another there were as many as 50 other manufacturers producing Hitchcock chairs, as the style came to be known generically. Today original Hitchcock chairs are often among the prized offerings at antique auctions—bringing bids of several hundred times their original cost—and stenciling, originally developed as an industrialized substitute for handwork, has been elevated to an American folk art.

Clockmaking was the next craft-based industry to be transformed by mass production in the early years of the nineteenth century. By 1840, American clocks were outselling British ones. A correspondent of the *Rochester Democrat* noted in a letter from Hartford, Connecticut, that 500,000 clocks were being manufactured annually in Hartford and that a quantity of clocks exported to Great Britain at an invoice price of $1.50 with a tariff of 20 percent had been seized in Liverpool on the ground that they were undervalued. They were released for sale only after evidence had been presented that they could, in fact, be manufactured at that low price and still provide a profit to the manufacturer. These were eight-day clocks, and since the item was unknown in England they were sold at auction for about $20 each—better than 1,000 percent over cost. These and many other American products that were manufactured to meet the demands of the middle class were produced in such quantities that they not only satisfied American markets but also found ready markets abroad, where their ingenuity and low cost gave them an undeniable appeal.

American pressed glass-
wares were invented to
provide a democratic
substitute for the far more
expensive blown glass.
However, as illustrated
by this compote, soon the
patterns in pressed glass
developed a character
of their own and could not
be cut on a wheel. Corn-
ing Museum of Glass,
Corning, N.Y.

In the early days of the
republic it was not unusual
for glass to be pressed
into designs with patriotic
themes, as in this plate.
Corning Museum of Glass.

Democratized elegance, it would appear, had just
as much appeal abroad as it had in the United
States—if the price was right. Even so, the general prejudice still favored foreign imports. Per-
capita imports almost doubled between 1830 and
1840, largely because some areas of American
manufacturing were not developing as fast as
others. As a result, the federal government instituted in 1842 a new structure of restrictive tariffs that virtually eliminated duty-free imports
by setting penalties (averaging 33 percent) on
all foreign goods. The effect of the new tariffs
on American manufacturing was startling. Within
10 years the production volumes of many items
tripled and quadrupled. Numerous patents were
issued for new machines and processes, as well
as new products, as it became evident that the
greatest market for manufactures lay in providing
products that could be bought and used by the
greatest number of people.

One area of nineteenth-century American enterprise that deserves particular attention, despite
the fact that it was to a degree outside the head-
long rush to industrialize, is represented by the
ennobling philosophy of service before self
expressed in the functional aesthetics of the
Shakers and their products. This communistic
theocracy of celibates, removed from the tempta-
tion of aristocratic pretension, emerged as a
reprise of the Puritans. The Shakers were dedicated to the dogma that clarity of mind, precision
of talent, and thrift of effort were the true paths
to glory. Through all of their buildings, furniture,
clothing, appliances, and farm products their
manifesto shines as a beacon for democratic
design: "[Let] it be plain and simple . . . unembel-
lished with any superfluities, which add nothing to
its goodness or durability." In an extraordinary
flowering of creative genius, the Shakers
invented and produced appliances, mechanical
devices, and accessories—many of them
patented. Some of their simpler products were
nests of oval boxes, flat corn brooms, and
wooden clothespins. They even came up with
a complete washing-machine system.

The Shakers endowed their furnishings with
eternal grace. Unhappily, their philosophy of form
seems to have taken root and flourished abroad

The extensive forests of young America stimulated the production of countless wooden products. "Fancy" or painted chairs by Lambert Hitchcock (1795–1852) were no exception. The chairs were much sought after by the general public as a symbol of domestic refinement. Index of American Design, National Gallery of Art.

Eli Terry, Seth Thomas, the Willard brothers, and other New England clockmakers adapted their craft to mass production and opened a rewarding new era in the clock industry. This shelf clock by Aaron Willard had a weight-driven movement that replaced the long pendulum. Museum of Fine Arts, Boston. Gift of Mrs. Mary D. B. Wilson in memory of Charles M. Wilson.

The Shaker slat-back chair represents the essence of the sect's philosophy. "Beauty," which they associated with worldly ornament, had no place in their lives—they considered it a waste of time and therefore morally repulsive. In compensation, however, their sense of form and proportion produced a quiet harmony that did not need embellishment. *Index of American Design, National Gallery of Art.*

Eating, like all other activities, was communal for
the Shakers. This pro-
duced the need for large
simple tables. Chairs and
other small furnishings
could be hung out of the
way on wall-mounted
strips of pegs. Andrews
and Andrews, Shaker
Furniture.

These oval boxes, made
with pine bases and tops
wrapped with a band of
thin steamed birch, were
products of the Shakers'
drive for perfection and
glorification of work. The
shaped ends and copper
pinning of the strips suggest ornament. Andrews
and Andrews, Shaker
Furniture.

The clothespin, as con-
ceived by the Shakers,
was first sawn in long
sections, then cut apart
and rounded to avoid
damage to clothes. Index
of American Design, Na-
tional Gallery of Art.

more than it has in the United States. The success of Shaker products was entirely in tune with
the seemingly insatiable public demand for
comfort and convenience. However, while their
vernacular products exhibited a spare elegance,
the American public, it seems, was still infatuated
with the aristocratic posture of English and continental styles and eagerly followed the fashions from abroad.

The *Journal of the Franklin Institute,* which
reported all American patents for many years,
shows a steady increase in those granted for
domestic products (washing machines, churns,
cooking stoves, and the like) relative to those
issued to mechanical devices, scientific apparatus, and machines. The pressure for federal
protection became so great that in 1836 Congress was obliged to revise its patent laws.
It reestablished a Commissioner of Patents (who
appointed examiners to review every application
in order to determine the priority of the invention
as well as its degree of novelty and utility), and
set the period of monopoly on patents for 14
years with provision for extending it for 7 more
years. This was changed in 1861 and now stands
at 17 years with no renewal permitted. To assist
the Patent Office in its new obligation, Congress
also voted to establish a library and a gallery
of models and specimens of manufactures and
the like. However, later in 1836 a disastrous fire
destroyed the Patent Office, and with it some
2,000 models, 9,000 drawings, and 230 books,
and other records. Although Congress promptly
appropriated $100,000 to rebuild the records and
asked all those holding patents to redeposit
their copies of issued patents and such other
evidence as they had in the Patent Office, many
valuable artifacts of American ingenuity were
gone, never to be recovered.

Less than a decade later the Smithsonian Institution was established in Washington as a result
of the bequest of James Smithson, an English
scholar and scientist, to the United States "for
the increase and diffusion of knowledge among
men." The Smithsonian has in principle become
the repository of American arts and sciences;
however, it provides only a modest record of the

The "castle" of the Smithsonian Institution, located on the mall in Washington, D.C. Author's collection.

vast contributions that the applied sciences and design have made to the United States. The main record is scattered in hundreds of small community and industrial museums and private collections across the country, little of it properly documented. Even the largest and most successful American companies have made little or no scholarly effort to properly document their progress and achievements. The United States has made only modest moves toward the establishment of a national collection of industry's great contribution to the national culture. Most of the collections that do exist are displayed as curios rather than as documents of the philosophy of design that they represent. Such scholarship as exists tends to be either transparent or opaque—seldom translucent.

By 1840 manufacturers had realized that what Bishop refers to as the "Polite, Fine and Ornamental Arts" had become an indispensible ingredient in the success of domestic furnishings. They realized that, although the consumer might not always understand the mechanism or construction of a manufactured product, he felt that he could always depend upon what his senses told him about it. They realized that the consumer would always seek to elevate his taste by purchasing fashionable products that reflected a higher level of aesthetic appreciation. A few manufacturers concentrated their efforts on out-and-out cultural products such as pianos, melodeons, and seraphines for music in the home and printing and daguerreotype processes for visual gratification. Many others, however, began to pay particular attention to the notion that artistic values applied to utilitarian manufactures might also increase their saleability.

Thus began the long, sometimes frustrating but always exciting game of style between consumer and manufacturer. The consumer seeks to advance his taste forward and upward, not only for his private satisfaction but also for the approbation of his fellow citizens. The manufacturer attempts to anticipate the consumer's often volatile preferences and seeks either to place before him the exact style of product that he desires and will purchase or to increase the consumer's dependence upon the aesthetic judgment of the

manufacturer. This contest continues unabated as another element in the American design ethic. Though it has resulted in many excesses and more than a few tragedies, it has served to hasten the transformation of abstract science into consumable technology and to nourish an indigenous American culture.

One characteristic of the American consumer that was pointed out at mid-century was that, because he expected an article to last for only a limited time, he was reluctant to pay a premium for higher quality. Nevertheless, in his conviction that industry and technology were constantly improving products under the pressure of open competition, he insisted on evidence of such improvement in the appearance of the manufactured product from year to year. No change implied that the product was falling behind.

In a curious way that is not normally understood by critics of American popular culture, manufacturing introduced a semblance of the quality of rarity that is normally associated with aristocratic products. The ever-increasing variety in patterns and detailing and the deliberate seasonal changes facilitated by mechanization made possible a kind of contrived uniqueness that could be presumed to allow for the expression of individual taste. That this often produced a tasteless, almost desperate melange of styles was not so important as the promise that machine-made culture need not result in aesthetic homogenization. More disturbing, however, was the fact that design was becoming such an important element in the success of manufactured products that it encouraged the imitation, if not the outright piracy, of those designs that were most successful in the marketplace. (Incidentally, whereas Dunlap in 1834 had anticipated today's definition of *design* as the plan of the whole and had noted that it could be applied to the solution of every problem from a necessity to a nicety, conventional thinking of his period associated design with the fine arts. Therefore, as the application of art and style to manufactures became important to their marketing, the word *design* was used to identify such decorative treatments as were applied superficially to the form and surface of a product—be it one normally expected to have a patterned or textured surface, such as a fabric, a wallpaper, or a floor covering, or be it a utilitarian appliance or machine.)

The reconstructed Patent Office of the United States had not, however, begun to issue protection for the ornamental design of products or for their distinctive appearance. It would seem that American manufacturers were too dependent upon imported inspiration to tolerate any interference with their practice of borrowing and putting out for sale any newly imported design, especially in the area of fabrics, as soon as it became or threatened to become popular. American mills, operating behind a shield of domestic tariffs, were so adept at copying foreign designs and so quick to get them to market that the best of the imports were quickly beaten out of their fair market.

Design piracy became such a problem in international trade that other countries developed legislation in an attempt to curtail it. England passed in 1839 "An Act to Secure to Proprietors of Design for Articles of Manufacture the Copyright of such Designs for a limited time." William Carpinael, a Londoner of the time, described the circumstances that brought on this first such act:

In many branches of our manufactures much money is annually expended in the designing and producing of patterns for which, formerly, no legal protection could be obtained; and it constantly happened that when a pattern was brought out in any department of manufacture, which was approved of by the public, other persons, engaged in the same trade, quickly copied the successful patterns, and having paid nothing for the design, the copyist could bring his articles into the market at a reduced price, thereby depriving the original proprietor of all reward. Thus the enterprising and talented manufacturer of integrity had less chance of success than those in the same trade, who, employing no skill or taste, were willing to depend on copying the production of others. (16, 3)

In 1843 the act was expanded to cover not only flat pattern designs but also "any article of manufacture, having reference to some purpose of utility, and that, whether it be for the whole

of such shape or configuration, or only for a part thereof." However, according to Carpinael, these acts did not apply to "any mechanical action, principle, contrivance, application or adaptation except insofar as these may be dependent upon, and inseparable from, the shape or configuration or the material of which the article may be composed."

The passage of the Design Act did not go unnoticed in the United States. Trade between the United States and England was so interlaced that any legal action taken by one country to protect the original design of its products must certainly lead to reciprocal moves on the part of the other. The United States Commissioner for Patents, therefore, proposed a similar law in his report to the Congress in 1841:

The justice and expediency of securing the exclusive benefit of new and original designs for articles of manufacture, both in the fine and useful arts, to the authors and proprietors thereof, for a limited time, are also respectfully presented for consideration. Other nations have granted this privilege, and it has afforded mutual satisfaction alike to the public and to individual applicants. . . . Competition among manufacturers for the latest patterns prompts to the highest effort to secure improvements, and calls out the inventive genius of our citizens. Such patterns are immediately pirated, at home and abroad. . . . If protection is given to designers, better patterns will, it is believed, be obtained since the impossibility of concealment at present forbids all expense that can be avoided. It may well be asked, if authors can so readily find protection in their labors, and inventors of the mechanical arts so easily secure a patent to reward their efforts, why should not discoverers of designs, the labor and expenditure of which may be far greater, have equal privileges afforded them? (136, 276)

As a result of this proposal, Congress passed, on August 29, 1842, an "Act to Promote the Progress of the Useful Arts" that permitted the Commissioner of Patents to issue design patents granting a limited monopoly to qualified applicants for their design concepts, including among other elements "any new and original shape or configuration of any article of manufacture not known or used by others before." (137, 352)

The *Journal of the Franklin Institute* for November 1842 acknowledged that the new act would give great impetus to the development of design in America, and in a subsequent issue began the regular publication of patents. In the first year, 1843, fourteen design patents were granted, the first for a bust of Robert Burns and the next three a design to be cast in metal as an ornament to surround stoves, a design for a floor oilcloth, and a figure for ingrain carpeting. That same year almost 500 patents were issued for other inventions. Within 10 years, the annual number of design patents had increased to 100 (compared with 2,000 mechanical patents). The annual number of design patents issued did not reach 2,000 until 1920, but a surge of interest in styling and industrial design more than doubled the number of design patents to more than 4,500 a year by the mid-1920s. On the eve of World War II the flood of design patents reached its maximum of 6,500 a year. Since then, as the effectiveness of existing design-patent law has been questioned, the annual number has fallen back to some 2,000. There have been repeated attempts to provide more effective design legislation; however, the number of manufacturers in the United States who depend upon borrowed design is still greater than those who would prefer to be originators. It must be pointed out that imitation tends to destroy initiative and may in the end make a slave of the imitator to the same degree that it makes a pauper out of the innovator.

Over the years great battles have been waged over product aesthetics. Whenever there is an unusual profit to be made through a desirable form, manufacturers find themselves locked in slashing conflict over its paternity. Because the patent commissioners are not permitted to pass judgment on the quality of a design (only on its presumed originality), designers are often forced to exaggerate the form of their expression so that its character will be inescapable to the commissioners. Federal law, therefore, encourages novelty and extravagance rather than simplicity and eloquence. A good design idea tends to be coarsened as it comes down the ladder of taste to a broader consumer base. In the end, the pure design idea is transformed into a bizarre caricature of its original self. Moreover, in its myopia, the U.S. Patent Office has carefully dis-

associated the form of an object from its utility, so that although form may indeed follow function, as Lamarck and others proposed, it is prevented by law from ever catching up.

The American public's infatuation with domestic refinement and its struggle to elevate its standard of living brought out not only manufacturers interested in feeding its desires, but also the professional tastemakers that Russell Lynes has written about so searchingly. By the middle of the nineteenth century these new professionals realized that the middle class hungered for guidance in its efforts to achieve a higher level of culture, and they established a unique position between the consumer and the producer. They took it as their mission to be informed about and sensitive to changes in the cultural climate and to transform their observations into salable commodities—personal counsel to wealthy patrons, private advice to manufacturers, and published guidance for an aesthetically insecure public. These counselors of culture have become an important part of the American design ethic, and often their self-fulfilling prophecies have had a profound influence on the quality and character of the American environment.

Andrew Jackson Downing (1815–1852), one of the first American tastemakers, developed the basic formula by which most operate. They begin by disparaging an existing fashion in order to clear the way for a new one that they can then praise as more honest than its predecessor and therefore more sensible and appropriate for those who wish to be up to the moment. Until Downing (originally a landscape architect) came along, the orderly classicism of the Federal style had dominated American domestic and public buildings and was considered eminently suitable for the young republic. However, Downing's sensitive mind caught the change in the air, and he attacked the "tasteless temples" as examples of architectural imperialism. He built his own house in the Elizabethan style, later to be known as Gothic Revival, on the ground that since every man's home was his castle it should look like one. Downing recommended the castellated Gothic for those who could afford it and a "rural" Gothic for the more modest homes of the middle class.

Andrew Jackson Downing's cottage, in the Rural Gothic style now referred to as "Carpenter Gothic," emulated in wood the stone tracery of the original Gothic. The sensible plan placed a bedroom on the first floor for the old and infirm and pushed the kitchen out the back to get the heat, moisture, and odors of food preparation out of the way. *A. J. Downing*, Cottage Residences, Rural Architecture and Landscape Gardening.

The Gothic impulse permeated the American environment as a rejection of the implied imperialism of the classical. This chair and stand combined the Gothic and rococo styles in a self-conscious yet not inelegant attempt at stylish comfort. Metropolitan Museum of Art. Edgar J. Kaufmann Charitable Foundation and Ronald S. Kane.

This whimsical chair by German immigrant George Hunzinger combines the designer-manufacturer's inventive mixture of Renaissance revival styles with the influence of the machine process. Metropolitan Museum of Art. Gift of Mrs. Florence Weyman.

This design for a library bookcase by Charles L. Eastlake, in the Gothic style that he championed, reflects his conviction that sturdy furniture with pegged joints and no artificial finishing was "honest" and should "suggest some fixed principles of taste for the popular guidance of those who are not accustomed to hear such principles defined." Reference 29.

His two books on landscape gardening and cottage residences proved invaluable to those who were waiting for aesthetic guidance. Downing proposed that Americans could not be happy with utilitarian design and that they would happily accept higher standards. He preached that the young democracy had to cultivate those arts that elevate and dignify the character. His own writings and those of others in the same vein—some illustrated by Alexander Jackson Davis (1803–1892), another prophet of culture—were accepted as gospel by an aspiring public and lent force to the race to keep up with the latest styles, not only in architecture and furnishings but also in industrial products.

After mid-century, the influence of the tastemakers increased all the more as steam-powered and belt-driven machines spewed out lathe turnings and jig-sawn gingerbread for homes and furnishings, molders cranked out iron castings for buildings, stoves, locomotives and furniture, and stampers and pressers produced metal and glass wares by the thousands. "The taste industry," as Lynes pointed out, had "gradually become essential to the operation of our American brand of capitalism." (61, 4)

In all likelihood the most influential voice of the era was that of the British tastemaker Charles Lock Eastlake, whose book *Hints on Household Taste,* published in England in 1868 and three years later in the United States, was hailed as a manifesto for reform in the decorative arts. True to type, Eastlake first condemned the volatile styles of the day as insincere and then preached a panacea of simplicity, humility, and economy in the design of objects to replace the wanton extravagance of revival styles. His doctrine promised an escape from the treadmill of style for anyone who had cultural courage and for any manufacturer who sensed a promise of fresh profit in Eastlake's concepts. Charles Perkins of Boston, who edited the American version of Eastlake's book, found in it a particular message for his countrymen because, as he wrote, "we borrow at second-hand and do not pretend to have a national taste. We take our architectural forms from England, our fashions from Paris, the patterns of our manufactures from all parts of

the world, and make nothing really original but trotting wagons and wooden clocks." (61, 104) However, Perkins too, by duplicating Eastlake's ideas for Americans, was contributing to the American belief that Europe must be the fountainhead of American culture.

Despite the tasteful romanticism of Downing and the moralistic persuasion of Eastlake and his American disciples, the great Centennial Exposition that capped this period failed to show genuine aesthetic progress and made it painfully evident that the country was still aesthetically immature, at least in matters of design.

Industrial Arts and
Good Taste

*An American home has become something more
than its original intent. It distracts the individual
too much from mankind at large, tempts him to
centre therein wealth, luxury, and every conceivable stimulus of personal ease, pride and
display. The tendency is to narrow his humanity, by putting it under bonds to vanity and
selfishness.*

James Jackson Jarves, 1864 (48, 251)

Every great international exposition is conceived
to demonstrate the progress that has been made
in the arts and the sciences. It becomes an invi-
tational tournament whereby the host nation challenges others to enter their accomplishments in
the lists for fame and glory. However, beneath its
cultural splendor and good will, it carries the arms
of economic competition. The great Centennial
Exhibition that was staged in Philadelphia in 1876
was no exception. Under a proclamation from
President Ulysses S. Grant in 1874, Secretary
of State Hamilton Fish invited other nations to participate in an exhibition "designed to commemorate the Declaration of the Independence of the
United States" with a "display of the results of Art
and Industry of all nations as will serve to illustrate the great advances attained, and the successes achieved, in the interest of Progress and
Civilization, during the century which will have
then closed." (59, 19)

Expositions often follow periods of national un-
rest or depression, and tend to allay the concerns
of the public about the present by reaffirming national purpose, recounting national achievements,
and parading glorious promises for the future. In
this case, two of the concerns to be allayed were
the bank panic of 1873 and the charges of political corruption in Grant's administration.

It was a stated goal of the Centennial Exposition (held at Philadelphia) to illustrate "the happy
mean . . . which combines the utility that serves
the body with the beauty that satisfies the mind,"
in that "artful living which can be provided by
manufactures." (82, 7) The addition of ornamentation to utilitarian products in order to make them
more pleasing to the eye was claimed to occupy

that middle ground between fine arts and industry
to be known in the future as industrial art.

The nearly 10 million Americans who visited
the Centennial Exposition were offered, besides
exhibits in the fine arts of painting and sculpture,
the promise that the industrial arts, in forms that
they had formerly associated with an aristocratic
style of living, might now be made available to
them at affordable prices. They applauded the
promise that American industry was now able
to compete successfully with the more expensive
foreign manufactures. Bishop wrote in 1868 that
American manufacturers such as Reed and Barton "familiarize the American people with forms
of beauty and elevate the standard of public
taste." According to Bishop, "An American artisan can now command exact copies of the choic-
est plate in the repertory of kings . . . the American people are being educated in taste and love
of the beautiful." (11, III, 331)

However, something sounded wrong from the
very beginning of the exposition, when Richard
Wagner, who had been commissioned to write a
centennial march for the opening, let it be known
that the best thing about his pompous work was
the $5,000 that the Americans had paid him for
it. Apparently a majority of the nations that had
accepted the invitation to participate had not sent
their best work in the fine and industrial arts.
The English felt duty-bound to do their best because of their former relations with the American
colonies and their present political and economic
ties. The Japanese, as a result of Perry's visit and
the subsequent treaty, went all out with the most
expensive installation; their exhibit launched a
general interest in Japanese arts that exerted
a powerful effect on American decorative tastes
and influenced the work of designers and architects for many years. But other countries apparently looked upon the invitation not so much as
an opportunity to honor the democracy, but rather
as a chance to sell to the aesthetically impoverished Americans those products that were not
of sufficient good taste to satisfy their own home
markets. It was argued by some that because
the Americans did not produce art of the highest
quality, they could have no appreciation for it.
The American taste establishment, feeling that

Opening day on the Grand
Plaza of the Centennial
Exposition in Philadelphia,
May 1876. The Memorial
Building was one of five
main structures erected
on 450 acres in Fairmount
Park. Reference 59.

wealthy and knowledgeable Americans who had traveled abroad extensively were quite familiar with foreign products and thus able to select the best *objets d'art* for their homes, resented the implication that American tastes were barbaric. They suggested that, at the very least, the works of high quality they might purchase at the exposition would "benefit the community in cultivating a correct taste and a higher standard of excellence in art." (82, 185)

The American industrial-art objects exhibited at the fair were generally considered to be little more than ostentatious echoes of European eclecticism. Although it is difficult through the haze of years to see much that is different between the stylistic presumptions of American and European products of the time, one may conclude that the mixture of Gothic and Grecian elements on a chandelier exhibited by the Cornelius Company of Baltimore leaves something to be desired. The Mason and Hamlin organ was clearly identified as an Eastlake organ even though Charles Eastlake himself discounted any blame: "I find American tradesmen continually advertising what they are pleased to call 'Eastlake' furniture, with the production of which I have had nothing whatever to do, and for the taste of which I should be very sorry to be considered responsible." (29, xxiv) A tea service and a swinging ice-water pitcher by Reed and Barton were baroque forms smothered with romantic ornament. The Bryant vase in silver designed by James H. Whitehouse, the chief artist of the Tiffany Company, was conceived in an extravagant allegorical mode to honor the American poet's affinity for nature. This not inelegant vase struggled to tell its story with apple blossoms, eglantine, amaranth, primrose, ivy, fringed gentian, water lilies, Indian corn, cotton plants, waterfowl, and bobolinks. The American masterpiece of the exposition was considered to be the Century Vase from the Gorham Company, which was designed by George G. Wilkinson and J. Pierpoint and required 2,000 ounces of silver to build. The exposition's largest industrial-art product, this vase paraded 100 years of American history. On the top stood "America holding up the olive branch of peace and the wreath of honor, summoning Europe, Asia and Africa to join with her

in the friendly rivalry with which she enters on the second century of her existence." (82, 56)

In retrospect, the cacophony of historicism in the industrial-arts section of the Centennial Exposition can be seen to have resulted in an aesthetic headache for the Americans that lasted beyond the end of the century. The capricious mix of styles that earned for the period such epithets as "Late Halloween" or "Early Awful" had a disastrous effect on the aesthetic quality of most American industrial-arts manufactures, despite the importation of samples, illustrations, machinery, designers, and artisans (or perhaps because of it). Walter Smith blamed both the manufacturers and the public: "The multitude desires quantity without regard to quality, and a manufacturer with the aid of his machine saws and lathes panders to this taste by turning out vast quantities of products loaded down with florid and cheap ornament." (82, 228) Smith questioned the judgment of manufacturers when it came to selecting products to be copied, and deplored that the manufacturer most often put his own taste above that of the artist he may have employed to help him. Smith reflected the popular dichotomy in that, while he felt machine-made products could not be as good as those that were made by hand, he praised the fact that the machine was bringing what had once been available only to the rich into the homes of the poorest citizens.

The affection of the Americans for mechanical devices was illustrated at the Centennial Exposition by a number of ingenious contraptions by which some articles of furniture could be made to serve more than one function. A bed that could be transformed into a sofa and a washstand that became a writing desk were justified on the ground that a young couple furnishing their first home needed such adaptability for economic reasons. Additional evidence of the American preoccupation with comfort and specialized convenience was shown by convertible chairs for invalids and other chairs designed especially for sewing, typing, barbershops, and trains.

However, the most technologically significant industrial-arts products shown at the exposition

The Mason and Hamlin company exhibited this organ in the Eastlake style. Its decoration was considered by Walter Smith to be both quiet and massive and "free from all the abortions in the shape of ornament with which many pretentious instruments are disfigured." Reference 82.

Smith considered this
"seventy-two-light chan-
delier of Lacquer gilt in
imitation of fine unalloyed
gold" to be one of the
masterpieces in his book
on the Centennial Expo-
sition. Reference 82.

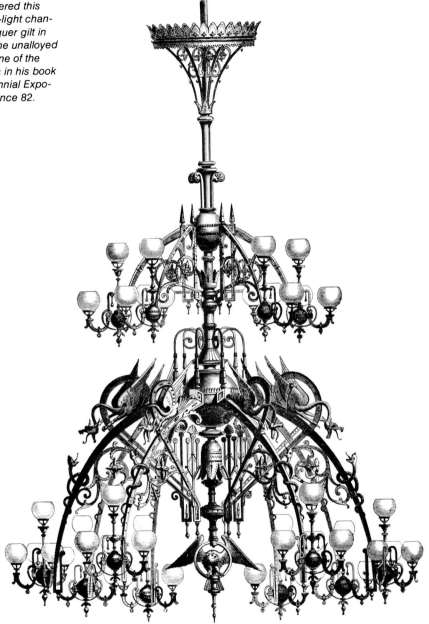

This silver-plated swinging icewater pitcher and tea service by the Reed and Barton Company was commended at the Centennial Exposition as being just as beautiful as its solid sterling equivalent with a cost that was "largely in favor of the plated ware." Reference 82.

The sterling silver William Cullen Bryant vase by Tiffany occupied a place of honor in the main building of the Centennial Exposition. It was the result of a subscription of $5,000 raised by friends of Bryant. For others it was available in a $500 electroplated copy. Reference 82.

The Gorham company's contribution to the Exposition was the solid sterling Century Vase, conceived to tell the story, in romantic allegory, of the republic that on its hundredth birthday commanded "the respect and admiration of the world." Reference 82.

The Aristocracy

were some pieces of furniture made by the bent-wood process that had been developed by Michael Thonet in the Rhineland some 40 years earlier. This type of furniture had first gained recognition in 1851 at the Crystal Palace in London, where a set of rosewood furniture was shown that demonstrated a basic commitment to the elimination of handwork such as carving and joinery in favor of knock-down construction for easy transport. In a way, the Thonet process was the nineteenth century's equivalent of the previous century's Windsor method of construction. By the time of the American centennial the Thonet Company's factories in Germany, Austria-Hungary, and Poland were producing more than 4,000 pieces a day, of which 85 percent were being sold abroad. The inventive rationality of the Thonet process predicted the future control of product form by manufacturing process rather than by artisanal whim. It resulted in an industrial aesthetic, as handsome as it was economical and useful, that served the democracy as well as it served the aristocracy. The Shakers, who had a small exhibit at the fair that is said to have influenced Scandinavian furniture design, were themselves intrigued enough by the Thonet process to produce several bentwood pieces of their own.

An article in the *London Art Journal,* cited by J. Leander Bishop in 1868 (11, III, 331), reflected the apparent insensitivity of the Americans to decoration: "The Anglo-American . . . seems the only nation in whom the love of ornament is not inherent. The Yankee whittles a stick but his cuttings never take a decorative form. . . . he is a utilitarian, not a decorator; he can invent an elegant sewing machine, but not a Jacquard loom; an electric telegraph, but not an embroidery machine." However, the Americans did find an interesting and even profitable accommodation between their interest in machines and their aspirations to become decorators in the fret- or jig-sawing process. The fret saw was a familiar hand tool that employed a C-shaped frame to hold a fine saw blade, generally made from a watch spring. Some unknown mechanic transformed it into a machine similar to the newly invented foot-pedaled sewing machine, thus making it suited

to female as well as male operators. Within five years after the machine was introduced, over 12,000 units had been sold. The version shown at Centennial Exposition, a continuous-band saw powered by the Corliss engine, "gave the business a great push." "Nothing in the exhibition of mechanical processes in Machinery Hall," according to *Harper's,* "had such a constant crowd of observers as one of these sawing-machines." (143, 533) It was estimated that within a year after the centennial 500,000 blades a month were being sold in the United States.

The American interest in industrial arts, given additional stimulus by the Exposition, was related to the notion that not only was the average home entitled to share the decorative furnishings of the American aristocracy, but further that each family could use its own talent and energy to ornament its own furnishings. "True decoration," announced the catalog of the School of Design at Cincinnati University, "may be said to be the beautifying of useful things. . . . Construction may be regarded as the peculiar province of men; to beautify is as naturally the province of women. The practical art department aims to instruct those who will be artisans and artists . . . those who will produce and those who will buy." It was generally presumed that interest in the industrial arts was the special province of cultured ladies; therefore, the schools of art and design that were founded within this rather narrow period (1867–1887) tended to concentrate on courses directed at them, ranging from nature drawing and design in the currently popular Queen Anne and Eastlake styles to practical instruction in pattern design, fret-sawing in wood and metal, wood-carving, needlework, and other similar skills. The avowed purpose of these courses was not only to enable one to "surround one's self with one's own creations and thus heighten the peculiar charm of the home" (143, 538), but also to enable the graduates to build reputations of commercial value.

Walter Smith, who had come to America from England in 1870 to serve as director of art and education in the state of Massachusetts, observed the following: "if the United States is to gain and maintain a place in the markets of the

Michael Thonet invented
not only the process for
making bentwood furni-
ture, but also the ma-
chines and tools to
manufacture it. He also
designed the factory that
housed the enterprise.
Walter Smith described
the furniture as "exceed-
ingly light and grace-
ful . . . especially adapted
to use in summer-houses,
where its lightness and
coolness make it agree-
able to the eye and
touch." Reference 82.

The foot-operated fret saw
had a particular appeal
to the Americans because
it combined the labor-
saving benefits of a
machine (however simple)
with rapid production of
ornamental products.
Furthermore, its associa-
tion with the sewing
machine made it accept-
able for women to operate.
Harper's Monthly, *March
1878.*

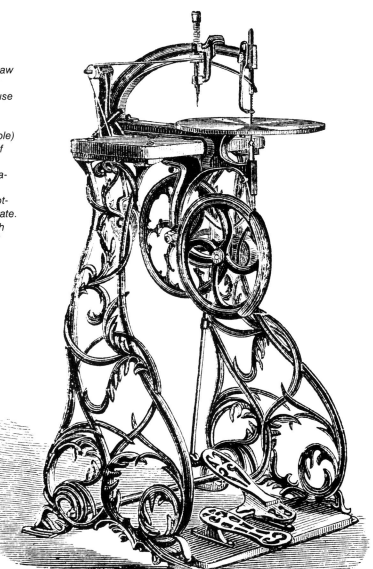

This fret-sawn clock housing in bird's-eye maple, ebony, mottled walnut, and ivory illustrates the capacity of the process for ornament. Harper's Monthly, *March 1878.*

The "Mohawk Dutchman wood-sawyer," a popular attraction at the Centennial Exposition. His versatile demonstrations on the bandsaw reflected the current interest in the sort of architectural ornament that has become known as "Carpenter Gothic." Reference 59.

President Ulysses S.
Grant and the Emperor
of Brazil started the great
Corliss steam engine that
served as the theme piece
of the Centennial Expo-
sition. The largest engine
ever built, it symbolized
the sweep toward indus-
trialization that was transforming the United States.
Reference 59.

Machinery Hall was a
great shed that carried
power from the Corliss
engine by way of pulleys
and belts to drive the
machinery that was being
demonstrated. Wilson, The
Masterpieces of the Cen-
tennial International Ex-
hibition, *volume III.*

The J. A. Fay company's
exhibition of woodworking
machinery. Some of the
machines were driven by
open belts, others by belts
beneath the floor. *Scientific American*, November
25, 1876.

world for manufactured articles of all description into which artistic design enters, . . . we must provide schools of art instruction and art museums in the great commercial and manufacturing centres of the country. And the sooner the manufacturers awaken to the necessity of this and act accordingly, the better it will be." Smith was convinced that the place to start was by teaching drawing in the public schools as a language and not as an art—as an instrument rather than a plaything. In fact, he felt that drawing should be taught before writing. Smith also said that no line could be drawn between the fine and the industrial arts, because artists often select utilitarian objects upon which to apply their art. "Ornamental art," he wrote, is "the fruition of industrial design" (82, 475), but, since utility constitutes more than half of the value of the object, "faithful service" must come first, then "graceful service." (82, 514) Smith acknowledged that there was already a school of design in Philadelphia, and must have known not only that the School of Design of the University of Cincinnati had a student exhibition in the Centennial Exposition but also that the Russian exhibit included examples from the Stroganof School of Technical Drawing, established in 1860 with the "view of forming an intelligent class of designers and ornamenters for the work of manufactories and industrial establishments."

The movement toward schools of art and design was well underway before Walter Smith came to America. Some were specialized private schools, such as the Minneapolis College of Art and Design (founded in 1867), the Art Institute of Chicago (1869), the Columbus College of Art (1870), and the Philadelphia College of Art (1876). Others were schools of art and/or design at private universities such as the University of Cincinnati (1870) and Syracuse University (1870). The Centennial Exposition provided the final incentive for the founding of the New York School of Decorative Arts (1877), the Rhode Island School of Design (1877), the Cleveland Institute of Art (1882), the Kansas City Art Institute (1882), and Brooklyn's Pratt Institute (1887). Recognition should also be given to the fact that the Maryland Institute of Art in Baltimore (1826) and the Moore College of Art in Philadelphia (1844), which preceded these schools by decades, were probably more concerned with the fine arts.

The exhibit of the School of Design of the University of Cincinnati in the Women's Pavilion at Philadelphia may have been typical of the work being done by the American schools in the industrial arts. Most of the 74 pieces that were shown were carved and fret-sawn products, ranging from a bedstead and a mantle to small inlaid boxes. In addition, there were designs for frescos and illuminations. The catalog lists 208 graduates of the school over an eight-year period, including some 50 women employed as lithographers, designers, sculptors, engravers, landscape painters, sign-painters and "stripers," architects, decorators, turners, and in other jobs. Those who did not find a place in industry apparently turned their talents to their own homes.

It is surprising that, despite the occasional mention of the application of decoration to utilitarian products and the rare use of the term *industrial design,* the general aesthetic reference at the Centennial Exposition was to domestic and public furnishings rather than machines. Yet many of the newly invented machines of industry, transportation, and farming displayed cast-iron structures that were decorative in form, with gold striping and colored embellishments. Such attempts to make strange new products palatable were criticized by the taste establishment as irreverent pretense, but machine products were developing an inherent aesthetic that would one day discard such costuming.

Industrialization and the
Good Life

*With the gradual accumulation of wealth, the
utilization of natural forces through the agency
of machinery, and the great improvements in the
means of transportation, the consuming power
of the masses has also greatly increased, and
many things which were only regarded as luxuries have come to be considered by even the
humblest in the light of necessities.*

David A. Wells, 1875 (142, 721)

The centerpiece of the Centennial Exposition
was the giant twin Corliss steam engine, which
was put into operation by President Grant at
the opening ceremony to power the miles of belts
and thousands of pulleys that ran the exhibited
machinery. In a portent of the future, this engine
was also used to show that steam power could
be used to generate electricity, even though the
electric light (Edison, 1879) and the electric
induction motor (Tesla, 1888) had yet to be
invented.

Exhibits in Machinery Hall showed the latest
refinements in the American System of Manufacture, demonstrating that a product designed
for machine production could be made substantially less expensively than, and as well as, the
best handmade product. The American Watch
Company of Waltham, Massachusetts exhibited
the process by which it was able to manufacture
a watch that could be sold for a dollar. The American System, originated by Eli Whitney in the
United States, was first applied to watchmaking
in 1850 by Aaron Lufkin Dennison and proved
to be a remarkable example of the application
of mass manufacture to a vernacular product.
It resulted in a near-perfect typeform for a watch
that was of high quality and yet could be made
available to everyone at a low cost. The following
principles of mass production were laid out by the
American Watch Company:

The product must at the onset be conceived to
be better than its competition in order to gain and
hold immediate market acceptance.

It must be designed especially for mass-production processes rather than hand manufacture.

Individual parts must be interchangeable, and
they and all operating systems must be pretested
to meet exacting predetermined standards of
endurance and performance.

Time, space, facilities, and capital must be sufficient to support the volume of work and to sustain
the organization until it becomes self-supporting.

Distribution methods, marketing structures,
and advertising programs must be commensurate
with projected production quantities.

Foreign observers were obliged to acknowledge
that America's progress in manufacturing technology would force them to change their own
methods. As one of the Swiss commissioners to
the United States wrote to his countrymen: "For
a long time we have heard here of an American
competition, without believing it. The skeptics,
and there were many of them, denied the possibility of a competition at once so rapid and so important. . . . I sincerely confess that I personally
had doubted that competition. But now I have
seen—and I have felt it—and am terrified by the
danger to which our industry is exposed." (210,
11) As a result of his expression of concern, the
Swiss were forced to improve their watchmaking
industry until it was once again in competitive
balance with that of the Americans.

It was evident that the design of products for
mass production would have to observe rules
of economy in materials and labor and in efficiency and dependability of service that would
result in a machine aesthetic distinct from that
of industrial-arts products. Wrote James Jackson
Jarves in 1864: "We do not disparage the mechanical arts. They are as honorable as they are
useful. Whenever our mechanics confine themselves to those utilitarian arts the knowledge of
which is their professional study, they make their
work as perfect of its kind as that of any other
people. But when they seek to superadd beauty,
a new principle comes into play, which requires
for its correct expression not only a knowledge
of aesthetic laws, but a profound conviction of
their value." (48, 233) In this Jarves was reacting
against a presumption that was as troublesome
to designers in his time as it is today: that a utilitarian object that is manufactured can be given
a cloak of aesthetic respectability with applied
ornamentation and stylistic mannerisms. He did
not doubt that "the American, while adhering
closely to his utilitarian and economical princi-

ples, has, unwittingly, in some objects to which his heart equally with his hand been devoted, developed a degree of beauty in them that no other nation equals." (49, 323) In this respect for a machine-made utilitarianism he was reflecting the position of moralists like Ruskin, who argued that art, like nature and life, must be sincere, and rationalists like Lamarck and Darwin, who proposed theories of organic evolution as a struggle of an organism against the environment to evolve from a lower to a higher state of being. The heart of the theory was the proposition that every product, natural or man-made, was in effect a typeform and acquired beauty to the degree to which it approached the perfection of its species. Jarves believed with Emerson that the objects that man makes acquire aesthetic value because they spring irresistably from the wants and ideas of a people. Horatio Greenough, before him, had propounded the same philosophy and had discounted the notion that the form of the products of the mechanics was either accidental or inexpressive: "No! It is the dearest of all styles! It costs the thought of men, much, very much thought, untiring investigation, ceaseless experiment. Its simplicity is not the simplicity of emptiness or of poverty, its simplicity is that of justness, I had almost said, of justice." (41, 172)

Despite the observations of Greenough, Jarves, Emerson, and a handful of other Americans, it was the Europeans who sensed that a new design ethic was emerging in American utilitarian products. It was European visitors to the Centennial Exposition who, by comparing American products with their own clumsier equivalents, were able to develop a rationalistic theory of design based on utility. For example, one German observer at the fair wrote of the vernacular products that were exhibited: "The hatchets, hoes, axes, hunting knives, wood knives, sugar cane knives, garden knives, etc., are of a variety and beauty which leave us speechless with admiration. . . . on every hand we are met with the results of serious research and we are astonished to see how much our own well known tools could be improved. . . . Again and again we see examples of how American industry in its progress breaks with all tradition and takes new paths which seem

The "dollar" pocket watch became a typeform for the product as well as an international symbol of the fact that American products could be made by machinery from inexpensive materials and yet were equal in functional quality to the more expensive foreign competition, and dramatically lower in cost. Collections of the Greenfield Village and the Henry Ford Museum, Dearborn, Michigan.

AMERICAN WATCHES,
MADE BY THE
AMERICAN WATCH COMPANY,
AT WALTHAM, MASS.

TO THE PUBLIC.

Attention is invited to the following statement of facts in regard to these watches, and some considerations why they should be preferred to those of foreign manufacture:

Their sale has been constantly on the increase ever since the business was commenced—thus proving *that they have grown into popular favor through their intrinsic merits.* As an evidence of the extent to which they have received the endorsement of the public, we may state, that upwards of Thirty Thousand of them are now in daily use in the United States, giving perfect satisfaction to their owners.

This result has been effected in the teeth of the most determined and violent opposition from the greater part of those in the Watch Importing Trade in the large cities, who have systematically used all their influence with their customers, to discourage their dealing in an article which threatened, by its superiority, to displace the foreign watch to a very large extent. Many of the Jewelers and Watchmakers of the interior, a large proportion of them foreigners, seconded the efforts of the Watch Importers, being persuaded by their counsels, and misled by a contracted and imperfect view of their own interests; by the fear of loss on their stock of imported watches, and the apprehension that their profits might be diminished through competition in a well-known domestic article, with other groundless prejudices, arising from a superficial inquiry into the subject. Notwithstanding this, however, the watches have steadily gained in the estimation of the people, the retailers have been constrained to keep them to supply the demand, and by degrees, we are happy to add, their prejudices and alarms are being dissipated.

Our peculiar system of making the different parts of each watch the exact counterpart of every other watch of the same series, leads to a uniformity in quality which can never be attained by the foreign process. If one of our watches is good all are good; whereas each foreign watch is only a probability by itself, depending upon the skill and fidelity of the particular workman who may happen to be employed upon it. In addition to these primary conditions of success, every watch issued by the Company is made of the most choice and enduring materials, carefully finished by the various processes to which they are subjected, and then put together, inspected and severely tested by the best workmen in the factory. Such has been the care with which these various duties have been performed, that out of the large number of watches sold, not more than a dozen or two have been returned to the Company for exchange, from any cause whatever.

Every watch is guaranteed by a guaranty that is good for something, and by parties that can be readily reached. Foreign watches, of the most inferior description, are often *fully guaranteed* by their makers, whom it is impossible to call to account under any circumstances.

American watches come to the consumer unburdened by the various expenses and profits incident to importation—the total of which, including custom-house duties, more than doubles the prime cost of the watch before it gets to the pocket of the ultimate owner. This consideration of itself should decide the question in our favor.

Every dollar diverted from the purchase of foreign watches is so much saved to the country; so much encouragement to home industry, and so much added to the public wealth. We do not ask a preference on these grounds, if our watches are not *better* for the money than the foreign.

To conclude—we claim that our watches are the best and most durable time-keepers in the world, besides being the cheapest; and we assert that a series of watches was never made that would show so little average variation from true time as those we have issued. In individual instances, their performance has been unsurpassed by anything recorded in the history of horology.

A descriptive pamphlet, containing full information and numerous certificates from well known individuals, may be had on application to the undersigned.

CAUTION.—As our watch is now extensively counterfeited by foreign manufacturers, we have to inform the public that no watch is of our production which is unaccompanied by a certificate of genuineness, bearing the number of the watch, and signed by our Treasurer, R. E. ROBBINS, or by our predecessors, Appleton, Tracy & Co.

As these watches are for sale by Jewelers generally throughout the Union, we do not solicit orders for single watches.

For the American Watch Company,

ROBBINS & APPLETON, WHOLESALE AGENTS, No. 182 BROADWAY, NEW YORK.

Despite the success of the "dollar" watch, the American Watch Company found it necessary to advertise the fact that multiple manufacturing increased, rather than decreased, the accuracy and dependability of its products. G. & D. Cook & Company's Illustrated Catalog of Carriages and Special Business Advertiser.

to us fantastic." (78, 169) Later in the nineteenth century the principle of form as the result of function would become the rallying doctrine for functionalism, and in the next century it would be degraded into a functionalistic dogma with all of the presumptions of truth claimed by the earlier prophets.

The general public was not concerned with such lofty notions as the relationship of function to form or the inherent aesthetic of manufactured objects — it was simply overwhelmed by the flood of affordable machine-made products that promised to improve material existence. People were flattered by the realization that inventors were exploring every advance in technology that might be of service to them and that manufacturers were catering to their needs and desires. Although at the Centennial Exposition most of the utilitarian products were relegated to the Wagon and Carriage Annex of the main building, the public searched out and absorbed with hungry eyes the precursors of today's major appliances and an exhaustive array of ingenious smaller conveniences for the home. Among the gadgets were apple corers, ridiculed some years earlier by the English as "the last comical vagary of the funny and awkward American cousin." Other devices shown, according to *Harper's,* included "the almond-peeler, pea and bean shellers, peach and cherry stoners, raisin-seeders, bread and cheese cutters, butter-workers, sausage grinders and stuffers, coffee-mills, corn-poppers, cream-freezers, dish-washers, egg-boilers, flour-sifters, flat-irons, knife-sharpeners and lemon-squeezers." The article went on: "Nor must the washing-machine, another strictly American notion, be disregarded. There are hundreds of patents. The typical forms are few; the variations on these forms are most amusingly numerous. The ins and outs of invention have been wonderfully diversified. . . ." (141, 374)

The dominant domestic appliance of the 1800s was the cast-iron cooking stove, whose typeform had been established earlier in the century by Hoxie (1812) and James (1815). Within a few years there were many manufacturers competing with one another not only in functional innovation but also in rococo decoration, which was to per-

sist well into the twentieth century. The Whitworth and Wallis report of 1854 on American manufactures documented the American stove's unique combination of technology, function, and aesthetics:

Much ingenuity is often displayed in the arrangement of the parts of a stove, and the adaptation of the decoration to strengthen and sustain those portions requiring the greatest amount of metal. Great efforts are made after novelty, alike in construction, ornamentation and in name, for every stove has a distinct title by which it is known in the market; and the euphony of some of these is often more amusing than appropriate. . . . There is, however, a wide field for a better style than as yet prevails, and as an absolute necessity exists for a certain amount of decoration of surface alike to strengthen the panels, sustain the angles, and hide defects in casting, which would be too apparent on a mere plane surface, the ornamentation adopted often partakes of the character of an excrescence rather than of a decorative adjunct. (76, 267)

(Although the first gas stove had already appeared on the market by 1850, several decades after illumination by gas had been demonstrated in Baltimore, fuel supply lines were still not extensive enough to threaten the dominance of the coal-fired stove for cooking and heating.)

The most significant domestic appliance to come out of the nineteenth century was the sewing machine. European inventors had already done their share when the American Walter Hunt invented the first practical lock-stitch machine in 1833. Although Hunt manufactured and sold his machines, he refused (like Benjamin Franklin and Oliver Evans before him) to obtain a patent. As a Quaker, he was concerned about the "economic morality" of his contraption for fear that its introduction would injure the "interests of hand sewers." (152, 36) It remained for Elias Howe, Jr., to invent and patent, in 1846, the basic design that became the mechanical typeform for the sewing machine. (Although Howe earned over a million dollars from the manufacture of his machine, it was Isaac Singer who achieved fame and whose name became virtually synonymous with "sewing machine.")

These mechanical drawings of apple corers, from Edward H. Knight's American Mechanical Dictionary *(1876) illustrate the nineteenth-century Americans' preoccupation with invention.*

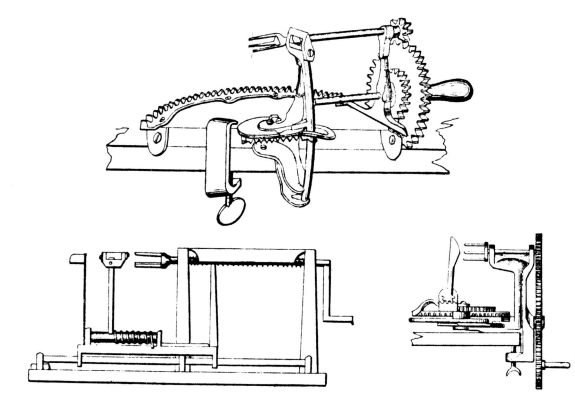

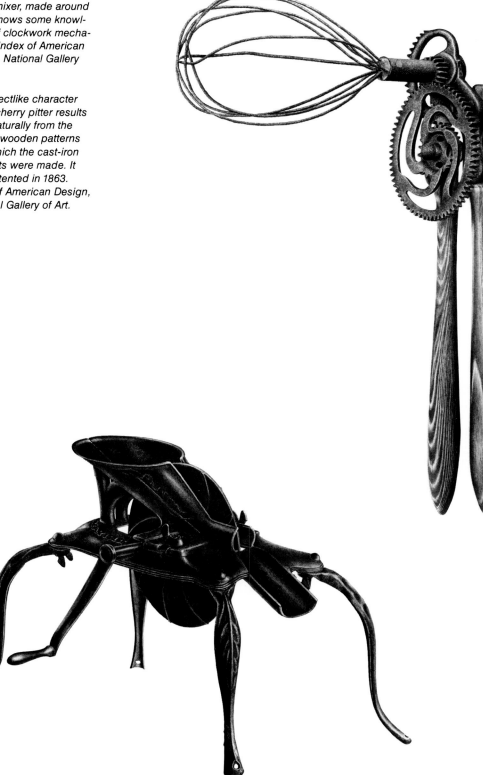

This mechanical cake-batter mixer, made around 1865, shows some knowledge of clockwork mechanisms. Index of American Design, National Gallery of Art.

The insectlike character of this cherry pitter results quite naturally from the carved wooden patterns from which the cast-iron elements were made. It was patented in 1863. Index of American Design, National Gallery of Art.

The "Uncle Sam" range,
promoted in the centennial
year, proposed to feed the
world. Its "beauty of
design" was said to be
unsurpassed. Library of
Congress.

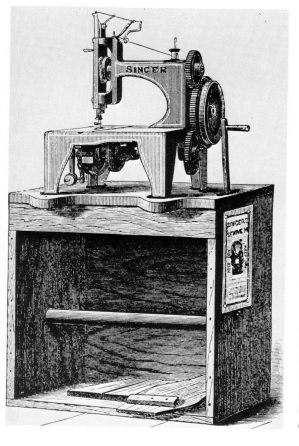

Isaac Singer's patent for a sewing machine, issued in 1851, was the first to include the final two features necessary for satisfactory operation: the presser foot and the ability to sew in a straight or a curved line. The crude shipping box served as a table for the machine and a housing for the treadle. Reference 9.

The sewing machine did more than any other product of the time to liberate women from domestic drudgery. At the same time, it industrialized the manufacture of garments, providing paid employment for thousands of women. In other words, it freed one class of women from interminable hours of slavery to needle and thread at the same time as it tied another class of women, mostly immigrants, to clothing factories. Good ready-made clothing became available to everyone at a cost of one-sixth of the time and energy required by hand sewing. And now that clothes were cheaper, the public was willing to exchange durability and service for style and fashion. This increased the demand for new clothing, encouraged further industrialization, and stimulated the distribution, promotion, and merchandising of the newly designed fashions that were created for each coming season. The Easter parade and the rush for back-to-school garments certainly owe their origins to the sewing machine.

The Singer Company, the major manufacturer of sewing machines, had its own pavilion at the Centennial Exposition celebrating the fact that, scarcely more than 25 years after Howe's patent, 600,000 machines a year were being manufactured (a third of them by Singer) and sold in the United States and abroad.

Even more significant than the sewing machine itself may have been the merchandising system that was devised by the Singer Company to put a sewing machine into every American home. The system was in perfect harmony with the need of an industry that is tuned to high-volume production to convince a cautious public to purchase its products. It introduced the three basic principles of mass marketing. First, *overcome public resistance.* Singer accomplished this by setting up demonstration agencies that employed women in order to overcome the prejudice of the time that women were too stupid to be trusted with machines. Second, *give something away.* The company announced that it would take in any machine or even a substantial part of a machine at a trade-in value of half of the purchase price of a new one. This was a remarkable offer, in light of the fact that Singer sewing machines sold at

retail for $100 at a time when the average Amer-
ican income was only $500 a year. And third,
spread the payments out. The Singer Company
introduced instalment purchasing on a national
scale. Though the size of the payments was
carefully adjusted to be within the buyer's ability
to pay, the company was able to charge an interest fee on the remaining debt, thus earning
a profit over and above the physical value of the
product. Thus the Singer Company, in the area
of domestic appliances, closed the operational
triangle of the American system of manufactures—mass production, mass distribution, and
mass consumption. This system demands that
the industrial designer be responsive to its three
measures, production, distribution, and consumption. One of these elements without the
others is impossible in the American economic
system.

The sewing machine was emblematic of the
explosive fervor of invention and the great rush
to industrialization that characterized the last
quarter of the nineteenth century, and it dramatically modified the lives and the economic and
social values of the American public. It served
to demonstrate, as the latest link in the modern
clothing system that had begun with Whitney's
cotton gin and Slater's mill, that every product
is only one component in a larger system of human service, and that any change in one component in the system will force a compensatory
modification in the other components. Though
it should be presumed that the components in
a system will strive constantly to achieve perfection, it is also true that any unchallenged down-
ward movement will pull the entire system down
to a lower level of quality. It is evident that human
beings, who are also components in every system that serves them, are monitors of those
systems. If they accept lower quality in any
product, their choice will draw the system to a
lower level; if they demand higher quality, they
will get it. Thus, it is fair to say that in a free and
open economy the consuming public gets what
it desires and deserves.

Alexis de Tocqueville had observed in *Democracy in America* that the Americans were content
to remain in a state of "accomplished mediocrity"

The Singer company
invited all 4,000 of its
employees to attend a
special celebration at its
building at the Centennial
Exposition. Reference 59.

This trade card was used to promote the Singer sewing machine, ca. 1885. The machine is still manufactured in Japan with the same sphinx decoration under the trade name "Mercedes," and in India under the name "Merritt." Author's collection.

The reverse of the same trade card showed a happy family at home with its Singer machine, and (most important) suggested that a young girl could operate the machine without fear. Author's collection.

This advertisement for the "Domestic" sewing machine (ca. 1885) showed newlyweds thrilled with their acquisition. Library of Congress.

because, in order to provide everyone rather than a select few with commodities, "the artisan in a democracy strives to invent methods which will enable him not only to work better, but cheaper and quicker; or, if he cannot succeed in that, to diminish the intrinsic qualities of the thing he makes, without rendering it wholly unfit for the use for which it is intended." "Thus," he wrote, "the democratic principle not only tends to direct the human mind to the useful arts, but it induces the artisan to produce with great rapidity a quantity of imperfect commodities, and the consumer to content himself with these commodities." (86, II, 58) There was no way that Tocqueville, perceptive as he was, could have foreseen that true mass production, with its computer-driven logic and automation, can only produce products of high quality if they are made in high volume and high speed in dehumanized manufacturing systems.

Tocqueville was convinced that a democracy under the guise of freedom of opportunity gave every man the opportunity to prey on his neighbor: "Handicraftsmen of democratic ages endeavor not only to bring their productions within the reach of the whole community, but they strive to give to all their commodities attractive qualities which they do not in reality possess. In the confusion of all ranks everyone hopes to appear what he is not, and makes great exertions to succeed in this object. This sentiment, indeed, which is but too natural to the heart of man, does not originate in the democratic principle; but that principle applies to material objects. To mimic virtue is of every age; but the hypocrisy of luxury belongs more particularly to the ages of democracy." (86, II, 59)

Tocqueville also noted the following: "Materialism is, amongst all nations, a dangerous disease of the human mind; but it is more especially to be dreaded amongst a democratic people, because it readily amalgamates with that vice which is most familiar to the heart under such circumstances. Democracy encourages a taste for physical gratification: this taste, if it becomes excessive, soon disposes men to believe that all is matter only; and materialism, in turn, hur-

ries them back with mad impatience to these same delights, such is the fatal circle within which democratic nations are driven round." (86, II, 173) Americans have been much abused for their presumed materialism. Though the criticism may not be without some justification, materialism may be taken as a token of their victory over a hostile environment and as a symbol of their emancipation from intractable political systems. It may be argued that Americans are advancing to a state of post-materialism, in which condition they are entitled by law and economics to the comfort and security made possible by modern technology. Thus, a commitment to provide the general public with what it needs and wants may be taken as a democratic base for large-scale production. In his introduction to *Democracy in America* John Stuart Mill called particular attention to the statement on the American preoccupation with public service: "It would seem as if every imagination in the United States were on the stretch to invent means of increasing the wealth and satisfying the wants of the people.

The best-informed inhabitants of each district constantly use their information to discover new truths which may augment the general prosperity; and, if they have made any such discoveries, they eagerly surrender them to the mass of the people. I have often seen Americans make great and real sacrifices to the public welfare . . . the free institutions which the inhabitants of the United States possess, and the political rights of which they make so much use, remind every citizen, and in a thousand ways, that he lives in society . . . Men attend to the interests of the public, first by necessity, afterwards by choice: what was intentional becomes an instinct; and by dint of working for the good of one's fellow citizens, the habit and taste of serving them is at length acquired." (86, II, 127)

The rapid industrialization of America was made possible by the expansion of a middle class that saw in industry a fountain of manufactured plenty that would elevate it to the level of culture and convenience of the aristocracy. The middle class claimed the right to own a completely detached home; the privilege of equipping the home with furnishings, furniture, appliances, and services reflecting owner's taste; the promise of an income that would permit the purchase of these products, with a little to spare; the assurance of enough leisure time to share experiences once reserved for the upper classes (vacation, travel, recreation, entertainment, and cultural advancement); and indulgence in style and fashion in personal care, costume, domestic furnishings, and other manifestations of a better life. That many of these pleasures were machine-made and promoted by cultural surrogates was not so important as the fact that their acquisition represented a step up the ladder of status and recognition.

In 1869 the sisters Catherine Beecher and Harriet Beecher Stowe published a small, somewhat idealized book, *The American Woman's Home,* as a "Guide to the Formation and Maintenance of Economical, Healthful, Beautiful and Christian Homes." It had three fundamental purposes: to strike a rather cautious blow for the liberation of women, to prove that a woman could run her home on scientific principles that were as valid as those that occupied her husband's professional time, and to demonstrate that the well-planned home of a middle-class family could be managed without the servants that were necessary in the larger homes of the rich. "At the present time," they wrote, "America is the only country where there is a class of women who may be described as ladies who do their own work. By a lady we mean a woman of education, cultivation and refinement, of liberal tastes and ideas, who, without any very material addition or changes, would be recognized as a lady in any circle of the Old World or New. The existence of such a class is a fact peculiar to American society, a plain result of the new principles involved in the doctrine of universal equality." (9, 307)

The Beecher sisters proposed that the status of women could be elevated if the public schools were to offer them Domestic Science courses on the same academic level as the professions of men. The result, they envisioned, would be homes designed to provide "modes of economising time, labor and expense by the close packing of conveniences." "By such methods," they

wrote, "small and economical houses can be made to secure most of the comforts and many of the refinements of large and expensive ones." (9, 463) They believed, some decades before Corbusier and others were given credit for the idea, that the true source of emancipation for women lay in the concept of the home as a domestic machine that was planned along scientific lines, with personal care, nutrition, education, and cultural needs to be served by organized methods. As a result inventors and manufacturers began to look upon the middle-class home as a lucrative market for labor-saving devices.

The concept of the home as a family-run food factory, essentially rural, with its icehouse, wellhouse, milkhouse, smokehouse, washhouse, and root cellar supplied by a vegetable garden, an arbor, an orchard, a cow barn, a pig pen, a chicken run, a corn crib, and a hay barn, was transformed into its urban equivalent by the invention of the glass preserve jar (Mason, 1858) and the tin can (Wilson, 1875). The ice-making machine (Gorrie, 1851) made possible the refrigerated railroad car (David, 1868), which carried fresh meat from Chicago, bananas from the Gulf of Mexico, and fruit and salmon from California to the rest of the country and led to a need for domestic refrigerators, which substantially reduced the housewife's trips to the market. In short order, the extension of water and sanitary systems to urban and near-urban homes encouraged the development of washing machines and sanitary appliances. And when gas lines were laid and electric wires strung to these same homes, the way was clear for domestic heating, cooking, and lighting appliances and an avalanche of other products for work and entertainment.

The dynamic expansion of all manufactures was entirely in harmony with a population increase from 25 to 75 million between 1850 and 1900. In large measure the increase was due to the great influx of immigrants who were attracted to the United States not only by the opportunity to work in the expanding industries but also by the hope of sharing in their cornucopia of products as members of the new middle class. It was natural that during this period the working population of the country would shift to a post-agricultural state, as fewer than half would be needed on the farms to produce food and fiber for all.

The inevitable result of this increase and shift in population was a general decline in the traditionally close relationship between the production and the consumption of products and services. As a result, a distinct consumer class began to emerge that was aware of the need to look to its own welfare. It began to recognize that it had an obligation to use its power to control the quality of its purchases and the conditions under which they were manufactured. The manufacturing establishment in turn realized that, with increasing consumer action and sharpening competition, it would have to plan and launch promotional campaigns to attract the sympathetic attention and the continued loyalty of its prospective segment of the consuming public. The costs of such advertising, as well as the rights of the buying public to select from a great variety of alternatives and to expect related services (such as home deliveries, orders by mail, and payment on an installment plan), would have to be borne by the consumers themselves. Experience has since demonstrated that there is a significant correlation between the volume of advertising and the high material standard of living in the common-market atmosphere of the United States.

The American practice of promoting products directly to the public gave rise in the 1860s to the development of advertising agencies such as N. W. Ayer and Son of Philadelphia and J. Walter Thompson of New York. By the time of the centennial, N. W. Ayer was able to claim that his agency could place advertising for a client in any newspaper in the United States and Canada.

Although the earliest newspaper advertisement in America appeared in 1704, advertising did not become an important factor in publishing until 1784, when the *Pennsylvania Packet and Advertiser* in Philadelphia was obliged to begin daily publication in order to meet the demands of prospective advertisers. This established the model for most newspapers that depend primarily on revenues from advertising, and in the next 100 years there was a phenomenal growth in the number of dailies (aided by the introduction of wood pulp into papermaking, by the modification of Hoe's rotary printing press to print directly from a roll of paper, and by the development of Bullock's high-speed web press). Printing costs dropped even more after 1885, when Mergenthaler introduced his linotype machine.

The growth of newspapers was paralleled by a rapid rise in the number of periodicals (from 1,200 in 1870 to some 2,400 in 1880) when they, too, began to carry advertising. Newspapers and periodicals are still the primary means of carrying detailed information about goods and services to the American public. The news that is printed and the essays that are presented often seem secondary to the dissemination of marketing information.

The increase in manufactures and the improvements in promotional media also stimulated the development of new methods of merchandising that essentially displaced the traditional over-the-counter sales of country and general stores. The Great Atlantic and Pacific Tea Company, which imported and retailed tea, coffee, and spices in the 1850s, grew into the A&P system of full-service grocery stores. Frank Woolworth's 5 and 10 cent stores (the first successful one opened in Lancaster, Pennsylvania, in 1879) sold notions over open counters that, in a preview of today's self-service stores, invited the customer to handle the merchandise prior to purchase.

Before industrialization, merchandise was carried to isolated villages and farmhouses by peddlers, and the availability of items was limited to what the peddler carried or what he remembered to bring the next time around. Then, after the Civil War, as the volume and the variety of products increased dramatically, the mail-order marketing system was born when Montgomery Ward launched his enterprise in 1872 with a single-sheet catalog. Before Montgomery Ward, sales by advertisement had been made only by occasional notices placed in magazines like *Allen's, People's Literary Companion,* or *Harpers Weekly.* Other retailers, among them R. H. Macy and John Wanamaker, had also tried to establish mail-order businesses, without success. R. W. Sears parlayed his mail-order watch business

The frontispiece for The American Woman's Home *pictured her as the central figure in three generations of domestic felicity. Reference 9.*

The idealized home of Catherine Beecher called for a central stove room on the first floor that could be closed off from the rest of the house. Fresh air was to be brought in to supply the main stove and the two Franklin stoves. Reference 9.

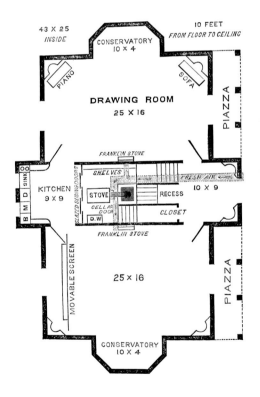

The Beecher sisters claimed that their stove could, with one ordinary coal-hod of anthracite, keep "running for twenty-four hours, keep seventeen gallons of water hot at all hours, bake pies and puddings in the warm closet, heat flat-irons under the back cover, boil tea-kettle and one pot under the front cover, bake bread in the oven, and cook a turkey in the tin roaster in front." Reference 9.

The Beecher kitchen work area, as well organized as a modern continuous-surface kitchen, provided for food preparation on one side and serving and cleaning up on the other. Reference 9.

Fig. 37.

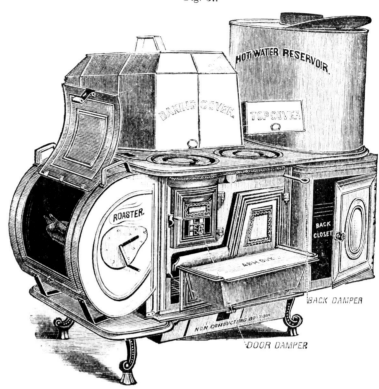

Mason's jar took domestic food preservation beyond the established methods. Mason was unable to protect his invention because it had been marketed publicly for more than a year before he filed for a patent. Collections of Greenfield Village and the Henry Ford Museum, Dearborn, Michigan.

J. A. Wilson invented a method by which meat could be preserved indefinitely. The tin can shape that he patented in 1875 continues to be made in the same form and has become generic to the corned beef product that it contains. U.S. Patent 161,848, April 6, 1875.

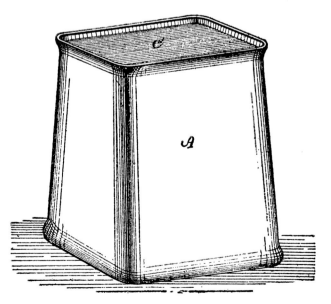

CATARACT
WASHING MACHINE.
—:o:—

DESCRIPTION.

It consists of a metal cylinder, with cleets on the inner surface, and an interior cylinder of wood, with cleets. There is a space of from six to eight inches between the two cylinders. One crank turns both cylinders at the same time in opposite directions, rapidly creating a suds, forcing the water through the clothes, and effectually removing the dirt.

ADVANTAGES.

This Machine dispenses entirely with the washboard. THE *action of the water* CLEANS THE CLOTHES, consequently there is NO WEAR OF FABRIC. The *saving of clothing*, and the *saving of time and labor*, are equally remarkable. The Machine is simple in construction and management,—a child can use it. It is well made, of galvanized iron, and is very durable. It will wash the *finest* as well as the coarsest fabrics,—a single small piece, or a quantity of clothing. For Flannels (usually the most difficult things for the laundress to manage), its operation is astonishing, as it thoroughly cleans them, with no possibility of shrinkage.

Prices.—No. 1, $12; No. 2, $14; No. 3, $16.
—o—

Machines can be seen in operation at No. **494 Broadway**, east side, above Broome St. Ladies and gentlemen are invited to call and examine it, *or, what is better,*

☞ Send your Dirty Clothes and test it. ☜

SULLIVAN & HYATT, Proprietors,
IMPORTERS AND DEALERS IN

American & Foreign Hardware,
54 BEEKMAN STREET, NEW YORK.

There were hundreds of patents issued over the years for a mechanically operated washing machine that would displace the scrubbing board. The Cataract washer advertised in G. & D. Cook & Company's Illustrated Catalog of Carriages and Special Business Advertiser of 1869 was one such machine.

The Home Washing Machine, manufactured in 1864, seems to have come closer to the typeform than many of the others. One geared crank worked a paddled wheel through the clothes, and another crank operated the wringer. Library of Congress.

The back roads and rural communities of America were served for many years by peddlers, who brought small domestic accessories, fabrics, and other products to the country housewife. Harper's Weekly.

The first Woolworth five-and-dime store, Lancaster, Pennsylvania, 1879, F. W. Woolworth Company.

of 1886 into his first catalog in 1891, and shortly afterwards joined with Roebuck to form the major mail-order company of the coming century.

Department stores, which have aptly been called the children of the machine age, could not have come into being without Otis's 1852 development of safe elevators to move people and products about, without improved means of communication and promotion, and without the cash register (Ritter, 1879), the adding machine (Burroughs, 1885), and the Addressograph (Duncan, 1892). R. H. Macy in New York and Marshall Field in Chicago opened this new chapter in the history of merchandising when they began to departmentalize their dry-goods stores in the 1860s, and Jordan Marsh of Boston transformed his wholesale house into a departmentalized retail store in 1861. The department stores became glittering showcases of plenty as slender cast-iron columns replaced thick masonry walls to create uncluttered interior spaces and large show windows to display the products of industrialization. In 1877 John Wanamaker convinced Thomas Edison to install an early electric lighting system in his store. (When the lights were turned on for the first time, it is said, crowds gathered in the street outside the store to watch it blow up.)

If one half of industrialization comprised industry itself, the appropriate machinery, and the transportation system needed to move raw materials and finished products, the other half was business, including the promotion and sale of the products and the management of the enterprise. Managers emerged as a distinct occupational class that provided the controls and records that were essential to manufacturing and distribution, and also the management of commodities and services that supported an industrial economy. Management found its place not at the factories (which were best located near transportation, raw materials, and energy sources), but rather in efficient and economical office buildings located in urban centers.

It was Chicago, more than New York, that provided the major opportunity for architects and builders to conceive structures that could be erected quickly in a form that was particularly suited to the needs of business. While the Eastern establishment was preoccupied with the Beaux Arts style, the younger architects of mid-America, perhaps stimulated by the need to rebuild Chicago quickly after its great fire of 1871, seemed to be determined to create an original American environmental character that was free of the stultifying traditions of the Old World. Yet, curiously enough, the Chicago World's Fair of 1893 would be cast in eclectic European styles, and one of the first great architectural competitions of the century to come would result in the Wrigley Building, which concealed its steel skeleton in a costume of skyscraper gothic. More than likely the historical garment served as a cultural disguise until an urban public could become familiar with the presence of the tall intruders in its midst.

Despite these aesthetic aberrations, the new architecture in Chicago was based upon the potential of steel. The Bessemer process of steel-making (1856) had reduced the cost of rolled steel sections to less than one-fifth of their original price and made it an ideal material for structures that would not only be less expensive but also lighter and easier and quicker to build. Moreover, it was discovered that a steel frame encased in concrete (also a new material to large buildings in America) would be essentially fireproof. The steel skeleton, pioneered by William le Baron Jenney to complete the top floors of the Home Insurance Building in Chicago (1885), became the primary method of construction not only for business buildings but also for railway terminals, hotels, factories, and public buildings. A framework of steel and iron covered with sheets of hammered copper was engineered by Gustav Eiffel for the Statue of Liberty, France's centennial gift to America. The second great monument of the nineteenth century, John Roebling's Brooklyn Bridge over the East River in New York (opened in 1883), depended upon strands of high-tensile steel wire to hold up the roadways. Steel, an engineered material that depended upon industrial methodology, had become essential for large structures, whose walls were transformed into curtains to become more of a burden to a structure than its support.

Advances in communication also benefited the business world. The Pony Express—a costly and unwieldy experiment that lasted only from April 1860 to October 1861—was abandoned not because of the hazards on its route between St. Joseph, Missouri, and Sacramento, California, but because telegraph service was extended from coast to coast. Five years later, Cyrus M. Field's transatlantic cable inaugurated dependable telegraph service between the United States and England. Alexander Graham Bell's telephone caught on quickly after it was demonstrated at the Centennial Exposition, and within a year there were 1,300 commercial telephones in use. Three years later 50,000 telephones were in service, providing businessmen with immediate voice communication from office to office and building to building. The typewriter (Sholes, 1867)—which had also been demonstrated at the Centennial Exposition by Philo Remington, who purchased the patent from Sholes—attracted enough attention as a business machine to open the male bastion of business to women, who were employed as secretary-typists. And then an avalanche of paper copies was set loose when Thomas Edison invented the mimeograph machine in 1876. Punch-card accounting (Hollerith, 1884) completed the rout of male stand-up bookkeepers. Even Waterman's fountain pen (1884), hailed at the time as a personalized writing machine, added to the technological revolution in business.

The inventions and the subsequent industrialization of the latter half of the nineteenth century would not have been possible without the combination of the assured aristocracy of the upper class with the assumed aristocracy of the middle class and the aspirations of the lower class. Only after a large enough base of consumption had been established was it possible to justify the construction and operation of complex interdependent product and service systems (such as those required for domestic, business, and industrial purposes) or of transportation systems. The national railroad systems that were developed during this period depended upon a national effort in which all classes participated and from which everyone would benefit. After the in-

vention of the air brake (Westinghouse, 1868) and the completion of the transcontinental link at Promontory Point in Utah (1869), thousands of travelers, rich and poor, abandoned the canal packets, the riverboats, and the prairie schooners for the railroads. Although the river steamboat did its best to hold onto its first place in American travel, its fate was sealed, as would be that of the railroads half a century later when scheduled air travel began. Such riverboats as were left were obliged to depend upon the romance of their appointments to justify survival. "This magnificent triumph of sculpturesque beauty," an advertisement for one proclaimed, "wedded to the highest grade of mechanical skill . . . must be as rococo in its upholsterings as a bed-chamber of Versailles—must gratify every sense, consult every taste, and meet every convenience." (151, 81) Nevertheless, steam locomotion, whether by water or land, depended upon all classes for support, and irrespective of the quality of the ride or the luxury of the accommodations the schedule for all was the same (much as in modern air travel, in which "coach" and first-class passengers arrive at the same time).

In urban and suburban areas, rail cars drawn by horses and mules were the major means of public transport. In 1884 Van Depoele developed the electric trolley car, which was successfully put into public service in 1886 on the Scranton Suburban Railway, but in 1890 the American cities were still using 105,000 horses and mules to pull 28,000 cars along 6,600 miles of track, helping to expand the cities as the people moved out to suburbs along the rail lines.

Signs of the coming automotive age were everywhere. The first commercially productive oil well was drilled near Titusville, Pennsylvania, in 1859 by Edwin L. Drake, and in 1872 George Brayton developed his first gasoline-driven engine. Newark, New Jersey, paved the first asphalt street in 1870. George Seldon filed the first American patent application for an automobile in 1879. He was wise enough, however, to manage somehow to delay its issuance until 1895, when the essential mechanical components of the vehicle and the environmental and economic components of the system were in place and he had

Only the torch-bearing arm of the Statue of Liberty, France's centennial gift to America, was finished in time for the Philadelphia exposition. Nevertheless, it was a major attraction. The designer was Frederick August Bertholde. Reference 59.

Proposals to span New York's East River with a bridge had been put forth since 1811, but it was not until the development of steel cables that the long-span wire suspension bridge became possible. The Brooklyn Bridge, designed by John Roebling, was opened to the public in 1883. Its gothic towers are often ignored in the drama of its modern wire webbing. Author's collection.

Introduced in 1876, the
telephone rapidly became
an essential part of everyday business. Its wires,
together with telegraph
and electrical cables,
spread across the streets
and the countryside.
Today such signs of
technological progress are
seen as visual pollution
and are being displaced
by underground or wire-
less systems. New York
Historical Society.

By the turn of the century, the telephone had become indispensible not only to the business world but also to personal life. Its typeform has changed twice since then and is now changing again. Library of Congress.

The typewriter opened the world of business to women. This basic configuration was patented by C. L. Sholes in 1867 and sold to Philo Remington in 1873. Only recently has this typeform left its plateau for the new forms made possible by the computer. Culver Pictures, Inc.

An advertisement for the
steamships of the People's
Evening Line, litho-
graphed by Currier and
Ives in 1878. The Drew
and the St. John are
shown passing each other
on the Hudson River be-
tween Albany and New
York. Courtesy of The
Mariners Museum, New-
port News, Va.

The saloons of the
steamships were open
spaces not unlike the
malls and hotel lobbies
of today. Their elaborate
appointments have yet
to be surpassed. Some
called them "floating
palaces." Courtesy of The
Mariners Museum, New-
port News, Va.

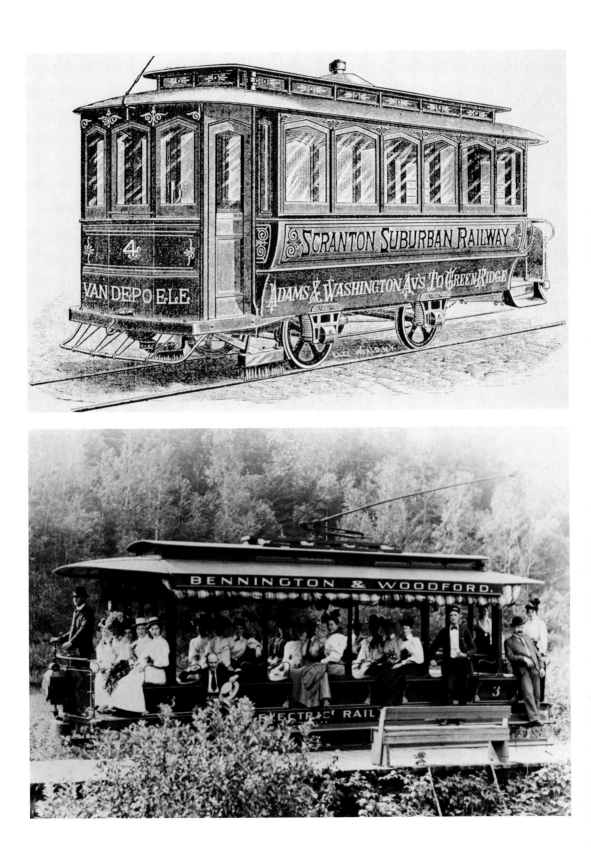

Charles J. Van Depoele's electric streetcar grew out of his experience in woodworking and electrical arc lighting. Power was supplied by a small trolley that was pulled along by the car on an overhead wire. Van Depoele's Scranton trolley ran 100 miles a day at 12 mph. Electrical World, December 1866.

The open trolley provided escape from the heat of a city summer—often to picnic grounds or amusement parks located at the city's edges. Robert L. Weichert collection.

George B. Selden's patent for a "road engine" was ignored by Henry Ford, who went ahead with his own automobile in apparent violation. By the time the courts held the patent valid, its period of monopoly had run out. U.S. Patent 549,160, November 5, 1895.

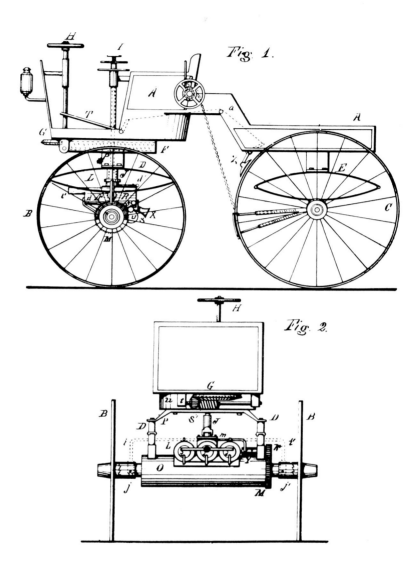

put together a coalition of American manufacturers that would hold a near monopoly over the manufacture of automobiles for more than a decade.

However, the automobile as the predominant means of transportation would have to wait until the next century, when a system of public highways would be built. For the time being, urban and nationwide railways dominated transportation and hastened the cultural homogenization of the Americans. The acquisition of land for the railroads, the control of the ores and forests, the manufacture of steel for the rails, and the building of the trains created an aristocracy of wealthy Easterners who would turn to the Beaux-Arts as evidence of their cultural ascendancy.

. . . to substitute the luxury of taste for the luxury of costliness; to teach that beauty does not imply elaboration or ornament; to employ only those forms and materials which make for simplicity, individuality and dignity of effect.

Gustav Stickley, 1901 (125)

Simplicity, and not the amount of money spent, is the foundation of all really effective decoration. In fact, money is frequently an absolute bar to good taste, for it leads to show and over-elaboration.

Louis C. Tiffany, 1910 (200)

The Centennial Exposition had promised that machines were capable of providing convenience and manufactured luxury for the citizens of the democracy. However, now that duplicated splendor was readily available, it was no longer good enough for everyone. Americans began to drift away from their egalitarian vision of society.

A new lower class was emerging, composed largely of those who had been liberated by the Civil War and the many new immigrants. For them the simplest forms of housing and domestic amenities were sufficient, at least for the moment, and in many cases vastly superior to what had been left behind.

There still remained the larger middle class, whose domestic felicities and cultural needs were being defined and directed by tastemakers and manufacturers. The influence of Eastlake, his American advocate Perkins, and others was evident in the displacement of the "honest" Gothic style by a capricious eclecticism of Gothic, rococo, Eastern, and Near Eastern details that were readily adapted to the band saw and the lathe. And where historical models were absent, inventive builders made up their own ornamental designs or purchased them from stock. This potpourri resulted in the so-called Queen Anne style in houses. (Queen Anne houses still survive in many American communities.)

The long-established upper class of Americans took its cultural course from the rediscovered heirlooms and the reawakened pride of its colo-

The natural fallout of the Beaux-Arts style were the so-called Queen Anne houses that combined Gothic finials, Palladian windows, French ornament, and Italian piazzas, all according to the architect's and the carpenter's imagination. American Architect and Building News, *September 16, 1882.*

An interior design described as "strongly marked with the Queen Anne feeling, although the mantle partakes somewhat of the Elizabethan period." Harper's Monthly, *August 1876.*

nial inheritance. The Centennial Exposition had revived an interest in colonial architecture and furnishings and helped to launch most of the major collections of early American artifacts. The Exposition contained three exhibits directly related to the colonial era: one of George Washington's clothing and equipment and a set of china that had belonged to Martha Washington, one of a hunter's camp, and (most important) a replica of a New England cabin built and furnished by the fair's Women's Commission. The cabin had a chair borrowed from the descendants of Massachusetts' first governor, John Endicott; John Alden's writing desk; and a teapot used by Lafayette. Of particular interest was a small spinning wheel that was presumed to have been brought over on the *Mayflower*—"Years ago it was thrown aside as useless," declared a contemporary review, "but when the Centennial movement began to extend its influence over the country, a Miss Tower took hold of it, burnished it up, and put it in condition to be operated on by her, much to the amusement of the visitors." (59, 87) The acquisition of such historic treasures not only preserved them for the nation but also provided for their owners a convenient retreat from the tasteless imposition of volatile foreign fashions. From the vantage point of a century, the colonial period came to be viewed as one of genteel elegance.

The several distinct styles of the time were blended into a generalized colonial style that continues in favor. American designers and manufacturers must take the colonial style into account when they are creating products for the national market, particularly domestic furnishings such as furniture, fabrics, silver, china, and glassware. Every manufacturer quickly learns that he must include a line of products in the colonial idiom in his catalog in order to attract and hold a sizable segment of consumers. Even the makers of modern products such as television sets, kitchen cabinets, clocks, and telephones are obliged to include at least one model in the cherished style. This affection for the comfortably stable historical style permeates the American design ethic and surfaces in preferences for certain colors, for homespun patterns in fabrics, and

for the natural textures of wood, stone, and brick in a wide variety of manufactured products.

Alongside the wealthy descendants of the colonial establishment (though not quite as secure in status) there appeared a *nouveau riche* upper class that, in an era of unrestricted competition, managed to capture the fortunes to be had from the harvest of America's animals, minerals, and timber, from the control of its transportation and communication systems, and from the management of its growing industrial and merchandising empires. Since they did not have a colonial heritage, the new rich looked to the aristocracies of Europe for social and cultural justification. They went abroad to acquire a touch of class, and they enrolled their children in the finishing schools of England and the continent. The wealthy American family trading daughter and dowry for a foreign title is a familiar theme, and theater and music abound with tales of the rich young American who marries the beautiful princess of an impoverished European dynasty.

Unlike the wealthy of the colonies, who were for the most part educated to culture, the new rich found it more convenient to employ collectors, architects, decorators, and other professional tastemakers to construct their copies of the palaces, mansions, chateaus, villas, and baronial estates of Europe, to manage their collections, and to adorn them with such examples of the fine and industrial arts as could be pried loose with American dollars. "For better or for worse," Wayne Andrews has written, "the millionaire was the American hero in the decades between Lincoln's assassination and Wilson's inauguration." (3, 152) Thorstein Veblen diagnosed the passion of the *nouveau riche* as "conspicuous consumption"—the notion that wealth must be masqueraded as beauty in order to gratify those who have it.

These were the decades, between the Civil War and the ratification of the Sixteenth Amendment to the Constitution, that established the idea of an income tax to make all Americans share equitably in the costs of operating their government. These were the years before the passage of the

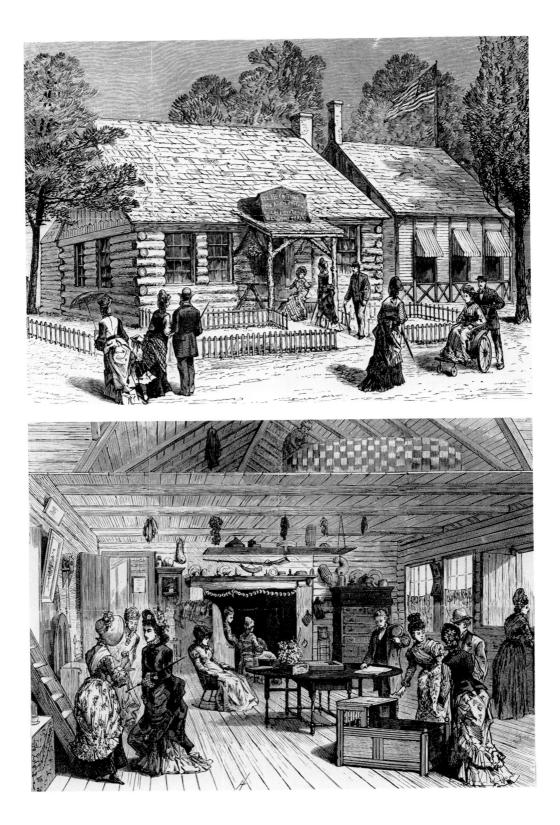

The log cabin built for the Centennial Exposition was considered to be in the style of architecture that characterized the settlers' cabins in colonial Vermont and Connecticut, but its many windows, neat picket fence, and self-consciously rustic entranceway were out of character with its predecessors. Reference 59.

The interior of the log cabin was called the "kitchen" by those who built it for the Centennial. In fact, it combined every domestic facility save sleeping space. Its furnishings were a mixture of antiques from the rustic to the refined, not unlike the blend that is called "Colonial" today. Reference 59.

Sherman Antitrust Act and the breakup of monopolies, when runaway fortunes could not be spent grandly enough and when extravagance at the top was presumed to be a socially appropriate means of ensuring a stable economy at the bottom. These were the Beaux-Arts years, when the wealthy Eastern establishment and its hired curators of eclectic taste imposed the fashionable French Renaissance, Italian Romanesque, and Graeco-Roman classical styles on the United States. The architects Richard Morris Hunt (1827–1895), Henry Hobson Richardson (1838–1886), and Charles Follen McKim (1847–1909) all had studied at the Ecole des Beaux-Arts in Paris, and their works represent the influence of the above three styles (respectively) on the American continent. In the eighteenth century, European architects had come to the United States to practice their professions. Now, in the nineteenth century, the flow ran the other way as Americans went abroad to study. (In the twentieth century, the flow would be reversed again as once again European architects would bring their talents to America.)

Richard Morris Hunt was the first American to study at the Ecole des Beaux-Arts. After some practical experience in Paris, he returned home with a comprehensive library of architecture books and hundreds of photographs of architectural details that undoubtedly served as a lode of ideas for the urban palaces and rural "cottages" he was to design for wealthy American clients. His first major commission, in 1879, was for a chateau built in the French Renaissance style on New York's Fifth Avenue for William K. Vanderbilt. Some 10 years later he built his first country residence, "Ochre Court" at Newport, for Robert Goelet. That was followed by "Belcourt" for the Belmonts and "Breakers" for Cornelius Vanderbilt II, also at Newport, and by "Biltmore," at Ashville, North Carolina, for William H. Vanderbilt. By the time Hunt was commissioned to design the administration building for the Columbian Exhibition in Chicago in 1893, he was president of the American Institute of Architects and dean of the profession that, some 35 years earlier, his successful lawsuit to be paid for professional services as an architect had helped to establish.

The W. K. Vanderbilt mansion at 52nd Street and Fifth Avenue in New York, designed by Richard Morris Hunt in the French Renaissance style with ornate gothic details, was cited at the time as the beginning of better things in industrial design. Architectural Record, *May 1908.*

Eastern architects who followed Hunt found a gold mine in Beaux-Arts. Their curious justification for the style in America is summed up in a quote from an album of the Vanderbilt town house: "The work will be the vision and image of a typical American residence, seized at the moment when the nation begins to have a taste of its own. . . ." (3, 157)

Henry Hobson Richardson's first important achievement after his return from Paris was to win a competition in 1872 for Trinity Church in Boston. His church, conceived in the strong Romanesque style for which he became famous, was later to be awarded first place in a poll taken in 1885 to select the ten best buildings in the United States. Four other buildings by Richardson were also among the first ten: the city hall at Albany, New York, the New York State Capitol at Albany, Sever Hall at Harvard University, and the town hall at North Easton, Massachusetts. His future work included the wholesale warehouse for Marshall Field in Chicago and the Allegheny County courthouse and jail in Pittsburgh, which was completed in 1886, the year of his untimely death. Whereas Hunt was proud of the authenticity of his designs, Richardson added to his work a personal rustic flavor and a feeling for inventive decoration that anticipated the Reformist designers who were to follow him. Louis Sullivan, an outspoken critic of Beaux-Arts, praised Richardson for his expression of the power and progress of the times. Richardson had been undoubtedly influenced by the Arts and Crafts movement and, in particular, by its strongest proponent, William Morris, whom he had met on a trip to England in 1882. As a result, he took an active part in the creation of the furnishings of his buildings, directing the craftsmen and designing some of the furniture himself.

Charles F. McKim and Stanford White (1853–1906) had both worked with Richardson before they formed an architectural partnership in 1877 after a sketching tour of pre-Revolution buildings along the New England coast. Their early commissions, for informal shingled country villas, employed natural materials in order to achieve a bucolic elegance. Then, in 1887, after the completion of a compound of houses in the Italian Renaissance style in New York for Henry Villard and his friends, they won the competition for the Boston Public Library, which became a showcase for the Beaux-Arts style in America and led to other commissions calling for architectural classicism, such as Pennsylvania Station in New York. The firm of McKim, Mead and White is primarily credited with stimulating the revival of the colonial style in residences in the United States. One of their first residences in the reborn style was the home of H. Taylor in Newport. This was followed by others in the calm but impressive dignity of Georgian and Palladian classicism. Other assignments included renovations of the Jefferson Rotunda at the University of Virginia and of the White House (under President Theodore Roosevelt).

It is difficult to understand why the eastern architects did not try to take advantage of the new building technologies that were being pioneered in Chicago. They preferred to leave the structure of the building to the contractor, while they busied themselves with selecting and detailing a historical style for the exterior. In this sense they were not unlike the contemporary stylists of some manufactured products, interiors, and buildings who provide elegant and dramatic drawings of the effect they wish to achieve but leave the execution to craftsmen and mechanics, with the result that the illustration may often be more honored than its embodiment.

In the midwest, Major William le Baron Jenney (1832–1907), who had studied engineering at the Ecole Centrale des Arts et Manufactures in Paris rather than the Ecole des Beaux-Arts, became the early focus of the Chicago school of technology-based architecture by designing and constructing the first high-rise building on an internal metal structure. Among his first employees were Daniel Burnham (1846–1912) and his partner-to-be John Wellborn Root (1850–1890), who had an engineering degree. Louis Sullivan (1856–1924) had studied for a time at the Massachusetts Institute of Technology and served a year as a junior draftsman in the office of Furness and Hewitt in Philadelphia before completing his apprenticeship in Jenney's office in 1874. From

Trinity Church in Boston was designed by H. H. Richardson in the strong Romanesque style for which he is known. However, the Romanesque revival attracted only a few adherents, chiefly in Boston and Chicago. Reprinted from Architectural Record, December 1956, copyright 1956 by McGraw-Hill, Inc. All rights reserved.

Like Latrobe before him, Richardson was involved in the furnishings as well as the structure of his buildings. This red oak chair, which he designed for the Court of Appeals in Albany, is based on his observations in Venice but also shows a unique, husky inventiveness. New York State Court of Appeals.

The H. Taylor house in
Newport, Rhode Island,
built in 1885–1886, is
judged as having inaugu-
rated the Colonial style
in homes. Its feeling for
classical order and serene
proportion offered a wel-
come relief from the ex-
uberance of the shingled
style then popular in the
resort homes of the rich.
Preservation Society of
Newport County.

The Home Insurance building, designed by Jenney and built in 1884–1885, was the first American building to incorporate the principles of the modern skyscraper. It was designed with a steel skeleton above the sixth floor and curtain walls that combined maximum durability and internal flexibility. F. A. Randall, The History of the Development of Building Construction in Chicago.

This photograph by Ralph
Marlowe Line captures the
strong rhythm of the pure
skyscraper in Sullivan's
department store for
Schlesinger-Meyer in
Chicago (1889–1904).
Courtesy of Ralph Mar-
lowe Line.

The first two floors of Sullivan's department store (now Carson, Pirie and Scott) are decorated with ornate cast-iron panels that provide a dramatic contrast to the severity of the upper floors. Courtesy of Ralph Marlowe Line.

there he went to Paris to complete his formal education in architecture at the Ecole des Beaux-Arts. Shortly after his return to Chicago he was hired by Dankmar Adler (1844–1900) to take charge of the office prior to being made a partner in 1881. And Frank Lloyd Wright had himself studied engineering for a year at the University of Wisconsin before refusing an offer to study at the Ecole des Beaux-Arts in order to apprentice in the office of Adler and Sullivan. These men are representative of the many Chicago-school architects who became immersed in Jenney's commitment to structural innovation and to a new aesthetic, based on revolutionary materials, that introduced industrial technology to building and promised to put form based upon method (and later upon function) ahead of form based upon predigested expression.

Most of the commissions for the Beaux-Arts architects came from wealthy easterners who were either erecting palaces for themselves or directing the building of churches, libraries, government buildings, and academic buildings along the East Coast. For such structures only the great styles of the past would suffice, whereas it seemed appropriate enough that the new factories, warehouses, merchandising centers, and office buildings of the American heartland could more properly be left to the more mechanical structures of midwestern architects. The professional difference between the eastern stylists and the midwestern mechanics lies, perhaps, in the fact that the eastern architects had less technical training and may, as a consequence, have been unable to understand, let alone share in, the structural innovations of their midwestern colleagues (who, some thought, lacked empathy for the historicism of their eastern colleagues). Lewis Mumford concludes that "the only contemporary style that manifested vitality, that of the mechanical age itself, was carefully kept out of the architect's training," and that "even the engineer, seeking confirmation from the esthete, bashfully hid his clean forms under melancholy iron foliage." (68, 201)

The climax of the Beaux-Arts movement in the United States was achieved in 1893 when the Columbian Exposition was staged in Chicago to commemorate the quatercentenary of Columbus's landing in the New World. Although some claimed that Daniel Burnham of Chicago, chief architect for the fair, sold out to the eastern establishment by accepting Beaux-Arts as the principal style for the fair buildings, it is more reasonable to assume that he was acutely conscious of his obligation to make the fair a truly national exposition and, therefore, in a gesture of good will decided that Beaux-Arts architects should be asked to design the exhibition buildings around the Court of Honor. Richard Morris Hunt was commissioned to set the aesthetic pace by designing the Administration Building that was to stand at the head of the Court in an exposition plan developed by Frederick Law Olmsted. Declared one of Hunt's assistants:

We have said that this edifice was intended to introduce the visitors to the Exposition into a new world. As they emerge from its east archway and enter the Court, they must, if possible, receive a memorable impression of architectural harmony on a vast scale. To this end the forums, basilicas, and baths of the Roman empire, the villas and gardens of the princes of the Italian Renaissance, the royal courtyards of the palaces of France and Spain, must yield to the architects, "in that new world which is the old," their rich inheritance of ordered beauty, to make possible the creation of a bright picture of civic splendor such as this great function of modern civilization would seem to require. (4, I, 131)

Most of the exhibition buildings were designed by midwestern architects who felt obliged to follow the Beaux-Arts theme set by the Court of Honor. Only Louis Sullivan refused to surrender to eastern eclecticism in his design for the Transportation Building. Though he was scored by a colleague for his experiment with impure vernacular architecture, Sullivan managed to conceive a structure that, while it carried overtones of the past, managed to stand apart from the others with its inventive ornament and polychromatic colors. Sullivan's was the only building at the Exposition to be honored by a European agency (the Comité Général des Arts Décoratifs de Paris), and that honor was for its ornament rather than its form. (In 1896, Sullivan would proclaim that "form follows function.") A few years later Sullivan carried his unique sense of ornament even farther in his designs for the Schlesinger Meyer department store (now Carson, Pirie and Scott) that was built in Chicago between 1899 and 1904.

Sullivan's Transportation Building was the only one at the fair to escape being painted white. It had originally been intended that all of the main buildings would be painted in the then-fashionable somber colors. However, when it became evident that the painting could not be finished in time, Frank D. Millet, the Chief of Decoration, decided to paint everything white. In desperation, he contrived a machine to spray buildings and statuary with a white lead and oil paint drawn from a barrel with a rubber hose connected to an air pump driven by an electric motor. To make certain that the purity of the white structures would be preserved, the burning of coal was banned around the main buildings.

The impact of the White City, as it was promptly christened by the public, was miraculous. Even though the structures were simply sheets of iron and timber coated with "staff" (a mixture of cement and plaster with jute fibers that had been invented by the French to be used for such purposes), the pristine classical facades and heroic sculptures glowed magically under the first major application of Edison's incandescent lamps and the sweeping beams of American and German searchlights. It was evident that the Americans of the midwest were more than ready for the Beaux-Arts style, which they had learned to associate with the affluence and the cultivated taste of the new millionaires. Moreover, the style was symbolic of the tides of imperialism that were sweeping the world at the time, including American interests in the Canal Zone, Puerto Rico, Hawaii, and the Philippines. Chicago's White City stimulated a preference for white houses. To this day, most houses built in the colonial English or Spanish style are painted white. (Of late, however, there seems to be a tendency toward the more somber colors that Downing advocated over a century ago in order that they might blend more harmoniously with the environment. This is also evident in the use of Cor-ten steel, bronze-anodized aluminum, and tinted or mirrored glass panels to reduce the stark presence of some large buildings.)

The industrial-arts exhibits at the Columbian Exposition were not particularly distinguished, despite their extravagance. Most of the masterpieces that were displayed represented a confusion of late Victorian presumption and overwrought eclecticism. The Magnolia vase in enameled silver, by the Tiffany Company, went even farther than most, with Toltec handles on a naturalistic body carried on a foot with overtones of Art Nouveau. Only the exhibit of the Rookwood Art Pottery Company showed signs that the saner Arts and Crafts movement had invaded the United States. (Women potters of Cincinnati had turned to thrown and slip-cast pottery before 1880 as background for their experiments in decoration in hope of developing an honorable occupation for their sex in industry. One of them, Maria Longworth Nichols, had founded Rookwood Pottery with this goal in mind. In time other art potteries were established, such as that at the Sophie Newcomb College for women at Tulane University in New Orleans, the Van Briggle pottery in Colorado Springs, and the Greuby pottery in Boston.) However, the exciting promises of the application of electricity to every area of life more than made up for the misguided extravagance of the industrial-arts exhibits and the spurious drama of the retrogressive architecture. Exhibits in the Machinery and Electricity buildings underscored the importance of electricity to the manufacturing and mining industries as well as its indispensible value to transportation and communication systems. Fairgoers were even more impressed by the possibility of entirely new appliances devoted to their comfort and convenience. They were intrigued by the use of electricity for cooling or heating and by the thought that cooking could now be done more "scientifically." Housekeepers were shown griddles, kettles, coffeepots, teapots, and other vessels with enameled bottoms in which copper wires were embedded to carry heating current. New, electric flatirons, with their upper portions composed of nonconducting substances, weighed only eight pounds—a far cry from the cast-iron sadirons. One company even showed an electric haircurler as a novelty. To look at the electrical exhibits, it was said, was like seeing a new world in which electricity would be adapted to serve a vast con-

Richard Morris Hunt's Administration Building for the Columbian Exposition was erected at the head of the Court of Honor. With the railway terminal behind, it served as the key entryway to the exposition and set the architectural theme. Reference 4.

The Columbian Fountain by Frederich MacMonnies, erected in a pool in front of the Administration Building, was claimed to have been inspired by a sketch by Columbus himself. In fact it was quite similar to the chief monumental fountain at the 1889 Paris fair. J. W. Buel, The Magic City.

Sullivan's Transportation
Building, with its "golden
doors," extravagant orig-
inal ornament, and rich
play of colors, stood out
in contrast to the white
palaces from the past that
surrounded it. Library
of Congress.

The Rookwood Art Pottery Company, founded as a respectable source of employment for women, became an important part of the Arts and Crafts movement away from western European historicism. New York Public Library.

Electricity transformed the Columbian Exposition into a fairyland at night. In the Electricity Building both Westinghouse and General Electric used it to illuminate their Beaux-Arts displays. The public's attention, however, was focused on the electric car, the electrified kitchen, and Edison's Kinetoscope (forerunner of the motion picture). Major Ben Truman, History of the World's Fair, 1893.

The display of windmills covered some five acres at the fair, demonstrating the use of wind energy not only as a source of power for water pumping and grinding but also for other farm and manufacturing tasks. Of particular interest at the time was the demonstration that wind power could be used to generate electricity. Chicago Historical Society.

sumer market. (In passing, it is interesting to note that in a field near the South Pond of the fair was exhibited an array of windmills that would have fascinated the energy-conscious American of today as much as it did the fairgoer of 1893.)

The developing preoccupation with transportation was evident in the combination of old and new forms that served the Columbian Exposition. Visitors came to the fairgrounds by steam train, electric streetcar, or steamboat. A moving sidewalk on Casino Pier took the steamboat passengers to land. The fair's canals and lagoon were plied by Italian gondolas and battery-operated launches. A Viking ship had sailed over from Norway, and replicas of Christopher Columbus's three ships had been towed from Cadiz. One could travel the length of the Midway Plaisance at relatively high speed on an elevated railway. A great Ferris wheel dominated the fairgrounds. Experimental automobiles had already appeared, but it was still a year before the first American patent for a gasoline-powered vehicle would be issued and a decade before the first powered flight at Kitty Hawk.

The Beaux-Arts movement, despite its late fling in the Columbian Exposition, was on its way out. Few new American "palaces" were built after the fair. Although the Beaux-Arts style persisted through the early years of the twentieth century until it was laid to rest in Mellon's National Gallery in Washington, its decline was sounded publicly by Louis Sullivan's discordant Transportation Building at the Chicago fair. Years later, Sullivan was still contemptuous of the "naked exhibition of charlatanry in the higher feudal and domineering culture" of the fair, claiming that it would take 50 years to be eradicated—as indeed it was when the rationalist movement in architecture was reimported to the United States. (83, 322)

The Columbian Exposition provided, unwittingly perhaps, the first dramatic exposure of the work of that small band of American reformists who were advocating rejection of slavish dependence on historical styles. The ornamental work of Louis Sullivan and the art wares of the Rookwood Art Pottery of Cincinnati were indications that at least

some Americans were in tune with the Arts and Crafts movement in England.

Although a wave of cultural uneasiness had been building since Thomas Carlyle had warned that the machine age had sacrificed means to ends and human to mechanical values, and despite John Ruskin's call for morality in the arts and William Morris's search for social justification for his craftwork, it was not until the 1880s that T. J. Cobden-Sanderson of England coined the title Arts and Crafts as a banner for the movement. To the more perceptive Americans, including Henry Hobson Richardson (who visited William Morris in 1882), the movement appeared to combine the asceticism of the Shakers with a sincere effort to give unique meaning to utilitarian products. In fact, there is an inescapable connection between the religious fervor of the Shakers and the religiosity of Morris and his disciples. Both believed that the glory of dedicated work and a return to naturalism would exorcise the devil of commercialization. However, whereas the Shakers, like Thonet, sought to eliminate all superfluous ornament from their product, relying instead on form itself as the expression of function, the English hoped to make the decoration of simple vernacular products once again worthy of the creative attention of artists. For both this represented a rejection of eclecticism and the Beaux-Arts in favor of forms that did not conceal their structure or their method of manufacture. And even when ornament was added, it was to be derived anew from the natural world.

It is generally assumed that the Arts and Crafts Movement was, in large measure, a protest against the reliance of industry upon machines for manufacture. This is not entirely true. The Shakers and Thonet welcomed manufacturing methods and, in fact, invented production machinery and processes that improved the quality of their work as well as increasing production. And England's William Morris (1834–1896), the sensitive poet, the versatile designer of environmental arts, and the consummate craftsman— often presumed to have been the archenemy of the machine—was to be credited later by Walter Gropius as the indispensable link between the world of art and the world of work. Morris considered Arts and Crafts to mean more than the making of products that were beautiful and utilitarian. He saw the maker in industry as a dehumanized part of the manufacturing process, robbed of initiative and trapped in the production of increasingly inferior products. Furthermore, he was convinced that industry's only goal was profit, and he resented the fact that it did not use machines at their highest potential but rather at the lowest level to manufacture the cheapest and meanest products that the public would tolerate. According to Watkinson it was not, contrary to what is usually said of Morris, the machine itself that he abhorred—"It was its use and abuse which moved his hatred: it was the spectacle of the machine (meaning irresponsible industry as a whole) destroying men while it made things, regardless of use or beauty." (89, 77) Morris made it clear that in his utopia the machine and the factory had a role to play if man could avoid being their slave. "Our epoch," Morris wrote, "has invented machines which would have appeared wild dreams to men of past ages, and of these machines we have as yet made no use." (66, 169)

Despite Morris's socialist leanings, the net effect of his work was to make the industrial arts even more exclusive. His products, as a form of highly personal expression, became even more removed from the public. It seems typical of the most sensitive designers that their products, while uncompromising in their perfection, are often denied for one reason or another to the masses they had originally presumed to serve. Often they take on a kind of cathartic humility that embraces moral and social values that salve the conscience of a select few rather than solve the needs of the many.

In the United States, however, the Arts and Crafts movement took a distinctly democratic turn as the middle class embraced its principles as a means of improving the cultural quality of the domestic environment. Americans organized Arts and Crafts societies in many communities to provide themselves with aesthetic guidance and workshops in which they could make their own domestic treasures. The prosperous business climate at the turn of the century encouraged a boom in

George W. Ferris, a bridge builder and engineer from Pittsburgh, built the giant amusement wheel in one year. It was 250 feet in diameter, carrying 36 cars with room for 60 people in each. The Ferris Wheel rivaled Eiffel's tower in Paris, and its descendants are now standard equipment for amusement parks. J. W. Buel, The Magic City.

home-building along the expanding lines of improved rail transportation. Furthermore, social changes had brought more educational opportunities to women, which stimulated their desire to pay more attention to the artistic quality of their homes. Educated women were emerging as a distinct class of consumers who demanded not only improved domestic appliances and services but also professional information about homemaking.

The *Ladies' Home Journal,* founded in 1883 by Cyrus H. K. Curtis, was one of the first magazines to address itself directly to the homes and furnishings of the average American. In 1889 the editorship of the magazine was assigned to Edward Bok (1863–1930), who accepted the philosophical goal of the Arts and Crafts movement to make the world a better and more beautiful place in which to live. Bok transformed the *Journal* into a very successful periodical dedicated to domestic elegance. He campaigned vigorously against bad taste by comparing in print good and bad taste in furniture and other domestic furnishings. Bok recognized that people shared a curiosity to look into others' homes—particularly the homes of those who were on a higher social and economic level, from whom they might learn about a more elegant way of living. Therefore he began to feature photographs of rooms in such homes for his readers. Bok also knew that most Americans could not afford to hire their own architects but had to rely on the self-serving taste of builders and contractors. As a result, he sought out the services of architects (including Ralph Adams Cram and Frank Lloyd Wright) who dared to risk the scorn of the architectural establishment by designing for *Journal* publications homes that could be built at a modest cost. The magazine sold such plans by mail for $5.00 a set, and it is said that hundreds of homes may have been built with their guidance.

In 1900 Bok commissioned William H. Bradley (1869–1962), who was already well known for his posters in the Art Nouveau style and as the owner of the successful Wayside Press, to assist him in developing the magazine's editorial policy. As part of their arrangement Bradley agreed to develop a series of illustrations for the *Journal*

of interiors and furnishings in the latest style. His designs showed that he was familiar with the work of the most influential Arts and Crafts leaders of England and Scotland. The series was presented in the magazine over a period of ten months, illustrating a cozy suburban home called the Bradley House. The *Journal* made this promise: "Mr. Bradley will design practically everything shown in the picture. It is not his hope that anyone will build a house completely as he designs it: he hopes rather to influence through individual suggestions—through pieces of furniture, draperies, fireplace accessories, wall-paper designs, all of which can be independently followed and detached from his entire scheme."

William Bradley's work for the *Ladies' Home Journal* signals the emergence of an American designer type—not from the profession of architecture, but rather from the field of illustration—as the source of product concepts that could be used to attract readers to a magazine at the same time as they would stimulate progress with promises of a better environment to come. The principle of incentive by illustration persists in the American design ethic as a means of generating interest in the potential of a new product and attracting the public's attention. Such "blue-sky thinking," as it is sometimes called, has fallen into question in recent years because the promotion played often drowned out the design. However, magazine illustration continues to have an important role in the introduction and acceptance of advances in technology.

Edward Bok's sensitivity to the desires and needs of the *Journal's* readers doubled the monthly circulation to over a million copies in a short time. By 1896 *House Beautiful,* with its first issue designed by Frank Lloyd Wright, had been founded to compete with the *Journal,* and within five years *House and Garden* had joined the competition for a growing market. That all three magazines are still in business is evidence that the interest of middle-class Americans in the quality of their domestic environment continues unabated.

*Frank Lloyd Wright's
house plans for the* Ladies'
Home Journal *were pub-
lished in Germany in 1912.
It was felt that the Ger-
mans were more hospita-
ble to new ideas at the
time than were the English
or the French—or the
Americans, for that matter.*
Architectural Record, *April
1912.*

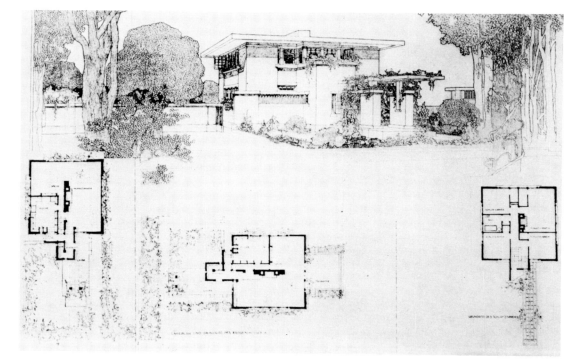

A Will Bradley design for
a reception/living room
and a hall, published in
the Ladies' Home Journal.
Bradley was called a
many-sided genius who
could design anything.
Although he was undoubt-
edly familiar with English
and American artist-
craftsmen, his own con-
cepts are rich, clear, and
consistent. New York
Public Library, General
Research Division, As-
tor, Lenox, and Tilde
Foundations.

Over the last decade of the nineteenth century and the first decade of the twentieth these three magazines, with the help of other more specialized publications, managed to stir up and sustain a strong interest in the American Arts and Crafts movement. Exhibitions of English products were held at prominent museums in Manhattan, Brooklyn, Newark, and Chicago, and some museums inaugurated permanent collections. Chicago was particularly attracted to the Arts and Crafts movement as a new wave in architecture and furnishings. Two English followers of William Morris, Charles R. Ashbee (1862–1942) and Walter Crane (1845–1915), were invited to lecture in Chicago, and in 1892 an exhibition of the work of Walter Crane was installed at the Art Institute. During the 1880s and 1890s fabrics, wallpapers, and furniture from Morris's workshops were shown and sold at Marshall Field's department store. After Morris died, a memorial room in his honor was dedicated at the Tobey Furniture Company in Chicago. (The Tobey Furniture Company had, incidentally, been formed by the brothers, Charles and Frank, in 1875 to manufacture ordinary commercial furniture. However, in 1888 they formed a subsidiary company, Tobey and Christiansen, to produce the more expensive furniture of the Chicago school's version of the Arts and Crafts movement. By 1890, after Charles Tobey's death, the Tobey and Christiansen Company, with William F. Christiansen, a Norwegian immigrant, as president, became the leading manufacturer of high-quality furniture in the United States. The company's products conveyed a characteristic Scandinavian flavor that anticipated the interest in Scandinavian furniture.)

Frank Lloyd Wright (1867–1959), who established his own architectural practice in 1893 after a five-year stint as chief draftsman for Adler and Sullivan in Chicago, had been deeply affected by the new spirit that was emerging in the products that people built for themselves. Like his former employers, Wright was determined to be free of the stultifying Beaux-Arts influence, and with other young architects [among them George W. Maher (1864–1926) and George Grant Elmslie (1871–1952)] he found a professional niche in the challenge of Arts and Crafts to the aesthetic establishment and its commitment to original and even radical ideas. Wright and his colleagues founded the Prairie school of architecture, which took its horizontal sweep from the flat midwestern landscape. They believed that a building should be "organic"—that it should grow easily from its site and that its materials should be those most natural to the area. Moreover, they proposed that a building, with its interiors and furnishings, should be conceived as an integral unit. Although the organic principle was not entirely new, it was admirably suited to the Arts and Crafts philosophy, and it promised once again to bring moral and aesthetic standards into balance. The young midwestern architects were convinced that a national style in architecture and products depended upon the ability of designers to meet and satisfy the daily needs and cultural desires of the Americans.

At first, Wright expressed a preference for the handmade products of Arts and Crafts enthusiasts, and he commissioned George Niedecken to produce many of his designs. Later, he came to prefer "clean-cut, straight-line forms that the machine can render far better than would be possible by hand." It is also evident that Wright was more than sympathetic with the taut geometry of Japanese architecture and furnishings. Even more likely than the influence of Japanese or handmade forms is that Wright's designs were conceived at the drawing board, reflecting the instruments of the designer rather than the workshop of the craftsman. "From the very beginning," he wrote, "my T-square and triangle were easy media of expression for my geometric sense of things." (91, 125) These tools, along with the compass, are the natural coefficient of the lineal, angular, and circular motions of the simpler machines of production, and thus constitute an inescapable link between the tools of design and those of industry. The same innocent empathy may very well have been behind the functionalist and modern styles that were to come. Wright was, even so, aware that geometric solutions were not necessarily sympathetic to the human figure: "I have been black and blue in some spot, somewhere, almost all my life from too intimate contact with my own furniture." (91, 145)

Frank Lloyd Wright's
house for D. D. Martin
in Buffalo, New York, illustrates the low silhouette
of the Prairie style. Its
long, unbroken horizontal
lines express its organic
quality of seeming to
be one with its site. Its
appearance is still fresh.
Architectural Record,
March 1908.

Wright's furniture for the
Martin house had vertical
lines, in seeming con-
tradiction to his commit-
ment to the horizontal. It
is mechanical in character
and as uncomfortable vi-
sually as it must have
been in fact. Architectural
Record, March 1908.

After 1900 the focus of American Arts and Crafts shifted to upstate New York as the New England societies abandoned the movement for a return to the colonial style of furnishings. A thriving furniture industry had developed at the western end of New York to take advantage of the hydroelectric power that was beginning to flow overland from Niagara Falls. Among the new enterprises that were attracted to settle in the area was that established by Elbert Hubbard (1856–1915) at East Aurora, near Buffalo. Hubbard gave up a career as a successful executive in the soap business in 1892 to visit Morris and his Kelmscott Press in England, and returned determined to awaken his countrymen to the philosophy and practicality of the Arts and Crafts movement. He sold his business partnership in order to establish the Roycroft Press to follow the Kelmscott example by producing handmade books in which all of the elements would be combined to create a work of art. He also began to publish a small monthly magazine, *The Philistine,* in order to broadcast his homilies on friendship and individuality in thought and expression and to promote the Arts and Crafts gospel. From publications bound in soft leather the Roycroft shops expanded laterally into bookbinding and then into the production of leather products. By 1896 Roycroft had expanded its operations to include a workshop to make the furniture for the enterprise. In obeisance to his commitment to Arts and Crafts, Hubbard ordered his furniture craftsmen (who were, for the most part, local carpenters) to make products that were simple and functional with an honest expression of their structure. When visitors to Roycroft expressed an interest in its simple handmade furniture, he expanded the woodworking workshop to manufacture products for sale. Although Hubbard made no claim to have originated the "Mission" style of furniture, it is evident that Roycroft furniture predated that of Stickley in following the "mission" principles of simplicity, durability, and quality. The title "mission" is now popularly applied to the plain machine-made furniture (usually of oak) that was made at the turn of the century.

The high-back Roycroft chair, although an early design, was produced during most of the Roycroft period. It is sturdy in proportion and a natural product of carpenters turned cabinetmakers. Princeton University Art Museum.

Products such as this lamp, made under the direction of Dard Hunter, illustrate the naive yet honest design that characterized the smaller arts at Roycroft. Mrs. George ScheideMantel.

Hubbard was not a designer, but his devotion to the philosophy of Arts and Crafts inspired those who provided the designs and samples of Roycroft products, including the cabinetmakers Albert Danner and Santiago Cadzow and the illustrator Victor Toothaker. The metal workshops that were established to provide hardware for furniture and other needs at Roycroft employed Karl Kipp, an Austrian immigrant, and later Darl Hunter, a young American who had studied at Ohio State University and who would later emigrate to Vienna.

In the years when Roycroft was at its peak—before Elbert Hubbard and his wife Alice perished on the *Lusitania* in 1915—many prominent scholars and designer-craftsmen found their way to East Aurora to visit, to lecture, and to compare ideas with Hubbard. After his death, his son Elbert II maintained Roycroft until the Depression closed the institution in 1938. The Roycroft "campus" has been revived, and it is now on the National Registry of Historic Places and open to visitors.

The Arts and Crafts movement in the United States reached its peak over the opening decade of the twentieth century in the work of Gustav Stickley and Louis Comfort Tiffany. These men represented opposite persuasions of the complex philosophy that drives the American design ethic—the first an asceticism guided by an inner revelation based on self-denial, and the second a sensuous preoccupation with a search for the gratification of aesthetic sensitivity.

Stickley (1857–1942) took the Arts and Crafts movement as an obligation to meet the needs of material existence in the simplest manner possible. Although trained originally as a stonemason, he was working as a furniture maker when he traveled to England and was caught up in the forms and philosophy of the movement. Stickley's particular affection for quarter-sawn oak as the ideal material for furniture, it is claimed, may have been fixed by a plain desk with rectilinear shapes by Arthur Heygate Mackmundo that was illustrated in the English magazine *International Studio* in 1897. When Stickley returned to the United States in 1898 he formed his own company in Eastwood, New York (now part of Syracuse), in order to put his own convictions about furniture into practice. The new Stickley Company displayed its first line of products at the 1900 furniture show at Grand Rapids, Michigan. This resulted in a distribution agreement with the Tobey Furniture Company that advertised it in the Chicago *Tribune* of October 7, 1900, simply as "The New Furniture." Stickley objected to the fact that his name was not mentioned and the agreement fell through. The line was as yet unrefined, attempting to combine Art Nouveau influences with the direct simplicity of Arts and Crafts. A year later, however, in 1901, after Stickley had changed the name of his company to Craftsman and adopted the motto "Als Ik Kan" (if I can), originally associated with the Flemish painter Jan Van Eyck, he displayed a major line of new furniture in what was to be known later as the Mission style at the Pan American Exposition in Buffalo. This showing established him as an important influence in Arts and Crafts. That same year Stickley began publication of *The Craftsman,* a monthly journal with Dr. Irene Sargent of Syracuse University as editor. Sargent was to help Stickley refine the principles that would guide his own work.

The determined simplicity of Stickley's early furniture supported his conviction that its "mission" was to create a "style beyond style," a "world of permanently valid forms." Stickley did not intend to create a new style, but "merely tried to make furniture which would be simple, durable, comfortable, and fitted for the place it was to occupy and the work it had to do." (64, 295) Stickley was not comfortable with having the term "Mission Oak" applied to his company's products, although he did recognize the publicity value in some association with the Southwest's Spanish missions.

Stickley's early Craftsman designs resulted in furniture that was heavy and somewhat over-scaled. Not until Harvey Ellis (1852–1904) joined the Stickley firm, in 1903, were its products re-scaled into lighter and more subtle forms. Ellis was a sensitive designer who had begun his career as a draftsman in Albany, New York, and had

worked for Richardson and for some midwestern architects before joining Stickley as an illustrator and editor of *The Craftsman.* He also designed a lighter line of furniture that included ornamental elements reflecting English and Scottish precedents. (This line was displayed, but apparently never entered full production.) Although Ellis was only with Stickley for a year before his death in 1904, he also designed several houses, including an Adirondack camp and a bungalow, that showed carefully worked-out details. The magazine offered plans and specifications for such houses. The bungalow style of residence, although it had originated on the West Coast and was primarily promoted by Henry L. Wilson of Chicago as cozy, quaint, and attractive, was adopted by Stickley because it was economical to build and was in accordance with his principles of honesty, simplicity, and usefulness.

The freshness and elegance Harry Ellis had brought to Craftsman products helped to increase sales to the point that Stickley decided to move *The Craftsman's* editorial and executive offices to New York City in 1905 and to allow franchised manufacturing across the country. (The parent factory remained in Syracuse.) However, the popularity of Mission furniture soon exceeded his ability to keep up with demand, and it was soon being widely copied by other companies. The plain and virtually indestructible furniture became standard in hotels, resorts, classrooms, and dormitories across the United States. Although Stickley tried to keep up with the avalanche of imitations by expanding his own operations, in 1915 he was driven into bankruptcy by the very success of his ideas. By 1916, when *The Craftsman* ceased publication, the Arts and Crafts movement in America had degenerated into what Siegfried Giedion has characterized as a hobby rather than a religion. Its aborted mission glows faintly in the surviving arts and crafts groups and shines somewhat more brightly in the connection between the Arts and Crafts philosophy and industrial design.

Louis Comfort Tiffany (1848–1933) saw in the Arts and Crafts movement an opportunity to grat-

Gustav Stickley's shop-mark combined a joiner's compass, his signature, and the motto Als ik kan ("If I can"). Its use in one of several variants marks a piece of furniture as being at the most desirable level of the Mission style. The Craftsman, 1904.

A catalog page from The Craftsman *showing a variety of Stickley furniture designed and manufactured for the living room. The page was illustrated in an article in the June 1908 issue of the* Architectural Record, *devoted to "Decorating and Furnishing the Country Home," which claimed that the "furniture known as the Mission style is now deservedly much in vogue."*

Gustav Stickley illustrated his version of the Morris "reclining" chair in the first issue of The Craftsman *in 1901, the same year he was granted a patent for its body. The chair was reproduced with very few changes as long as he was in business. Jordan-Volpe Gallery.*

This settle by Gustav Stickley was one of his most successful products. It combined soft leather cushions with a rectilinear structure that showed the grain of oak to advantage. Courtesy of Art Institute of Chicago.

ify his own desires, as well as those of other humans, for extravagant beauty in the luxuries of life. As the son of Charles Lewis Tiffany (1812–1902), founder of Tiffany Jewelry and Silversmiths in New York, the younger Tiffany was able to devote his early years to the study of art and to exhibit paintings at the National Academy of Design in 1867 and the Centennial Exposition in 1876. However, the philosophy of Ruskin and Morris convinced him that the fine arts of Western culture were too limiting, and he went abroad to study Islamic, Near Eastern, and Oriental art. Upon his return to the United States, Tiffany worked with John LaFarge long enough to learn the art and craft of decorative glasswork. In 1879 he founded, with others, the company of Louis C. Tiffany and Associates, which was to have a strong impact on the decorative arts. The work of the firm included a broad range of decorative arts, such as textiles, wallpapers, and the glass mosaics, tiles, and windows for which the firm became best known. Their most important commission was redecorating the White House in 1882–1883, during the Arthur administration. Tiffany produced an opalescent glass screen that was one of the executive mansion's main attractions until it was destroyed in 1904 on the order of President Theodore Roosevelt, whose many interests did not, unfortunately, include the arts.

In 1893, after several intermediate changes in the structure of his business, Tiffany had turned his efforts exclusively to glass and established his own glassworks at Corona, Long Island. Among the many excellent glass blowers and workers that he employed to execute his ideas was Arthur Nash, an English émigré, who is credited by some with having invented the distinctive "Favrile" glass in brilliant iridescent colors and rich textures for which the firm became famous. With Nash's help and that of many other artist-craftsmen, Tiffany was able to move away from the restrained forms and ornament associated with the Arts and Crafts movement in America toward the romantically daring Art Nouveau fashion that was beginning to entrance Europeans. The Art Nouveau style flared only briefly in the United States. Its exotic spirit is best re-

flected in Tiffany's rich "Favrile" glass wares and decorative accessories—especially the lampshades, which were based upon his imaginative interpretations of natural forms.

Louis Tiffany may properly be credited with having played a useful role in the sweeping popularity of the Art Nouveau style in Europe chiefly as a result of his friendship with the German entrepreneur Samuel Bing (1838–1905). Tiffany visited Bing's decorative-arts shop in Paris several times in the 1880s, and later he would make Bing his exclusive European distributor. When, in 1895, the oriental and orientally inspired decorative products that were the mainstays of Bing's business became hard for him to get, he shifted his focus to include work in the exciting new style and renamed his shop L'Art Nouveau, because, as he said, "we must seek the spark of new life beneath the ashes of older systems." To stock his new venture Bing sought out the most daring designer-craftsmen in the new style—including Tiffany, who sent him the best examples from his just-opened Corona glassworks. Others included Eugene Grasset (1841–1917), a Swiss-born Frenchman whose graphic designs were already well known in the United States through his covers for *Harper's Bazaar* and other American magazines; Emile Galle (1846–1904), the most innovative French glassmaker of the period; and René Lalique (1860–1945), an imaginative French jeweler who was to be an important force in the Art Moderne movement and whose glass perfume containers for Coty were to be much more valued than their contents. Henry Van de Velde (1863–1957), a Belgian painter-designer-architect, also contributed to Bing's shop by designing four rooms and furniture. Van de Velde and other Belgian, Dutch, and French artists and designers made up a group known as *Les Vingt*. They absorbed the principles of the English Arts and Crafts movement and the French Symbolists and transformed them into the Art Nouveau style.

Although the name Art Nouveau was French, the style was not particularly French in character. In fact, to the French it was known as *Le Style Anglais* in deference to its origins in the English Arts and Crafts movement. In Italy, where the

A black-and-white photograph cannot capture the rich color nuances of Tiffany's "Favrile" glasswares. These pieces show a variety of influences, from the Near East and Art Nouveau to romantic antiquity. Metropolitan Museum of Art.

The Wisteria Lamp, made by Tiffany around 1900 for Mrs. Curtis Freshel according to her original design, is probably the best of his lamps. It captures a spirit of naturalism without compromising the romanticism of Art Nouveau. Chrysler Museum; gift of Walter Chrysler, Jr.

style never really took hold, it was known as *Lo Stile Liberty,* since it was exhibited and sold in the famous shop that had been opened in 1875 on Regent Street in London by Sir Lazenby Liberty (1843–1917) on the advice of William Morris. The German equivalent of Art Nouveau was known originally as *Neue Stil.* However, over the years the term *Jugendstil* has come into common use to describe the style that began with the English Arts and Crafts and the search for new expression by the French and the Belgians. *Jugendstil* set its own course toward a technology-based aesthetic that found its outlet in the Deutscher Werkbund and the Wiener Werkstatte. In Austria the leading force for the new style was Otto Wagner (1841–1918) and his former student Josef Hoffman (1870–1955), who had the courage to break away from the visual-arts establishment in a movement to be known as the *Sezession.* In 1903 Hoffman established the Wiener Werkstatte, which found its own unique position between the Arts and Crafts and Art Nouveau philosophies and developed a particularly strong affinity with the Glasgow Arts and Crafts group led by Charles Rennie MacIntosh (1868–1928) and Herbert MacNair (1868–1955) and their wives, the sisters Mary and Frances MacDonald.

After the first decade of the twentieth century the Art Nouveau style lost vitality as its forms became stereotyped and began to be replaced by the more geometric forms of the Scottish, Austrian, and German designers. In the United States the works of Stickley and Tiffany influenced other designers and manufacturers and were echoed in products that were sold at every price level. After three quarters of a century, the strongest furniture fashions are again dependent upon the squared-off oak sections of Stickley (although the construction is now a laminated ghost of its original solid self), and Tiffany-style lampshades and windows are available in acrylic reincarnations.

Industrial Arts and the
Arts of Manufacture

We fabricate objects lighter and stronger than those of other nations, but we pay little attention to outsiders and have almost no consideration whatever for the probable prejudices and tastes of the people to whom we hope to sell.

An American visitor to the Paris Exposition, 1900 (162)

Americans at the turn of the century seemed to be preoccupied with the promises of an expanding technology, and in particular with the commercial exploitation of new energy sources. As petroleum and electricity became universally available in cheap and seemingly limitless supply they stimulated a frenzy of inventions based upon gasoline engines, electric motors, and heating coils.

The Pan-American Exposition, staged at Buffalo, New York, in 1901 to promote the social and economic interests of the countries of the western hemisphere, became a paean to electricity, which illuminated its buildings, its theme tower, and its fountains with energy drawn from the world's largest power plant at nearby Niagara Falls. The exhibits, ranging from manufactured products to Arts and Crafts objects, suggested to some an important change in the lives of middle-class Americans, with machines to make life easy and the application of art to everyday objects to make it more pleasant.

The living quarters of the home—its parlors, dining room, and bedrooms—had long been considered the realm of the industrial arts and dependent upon either the industrial arts or the Arts and Crafts for style. Now the kitchen, the bathroom, and the laundry, where function was held to be more important than fashion, looked to technology and science for assistance. The time-and-motion studies of Taylor and Galbraith, which had already demonstrated that labor in industry could be made more efficient through planning, stimulated women like Christine Frederick and Katharine C. Budd to propose that rational analysis could liberate women from household labor. Mechanical slaves, many of them powered by electricity, it was proposed, could replace servants in the home.

One class of electrified products for the home was based upon the generation of heat (and light) by the resistance of alloy wire to electric current. The most dramatic application of this principle was Edison's incandescent lamp (1879); other applications included the electric flatiron (Seeley, 1882), the electric stove (Hadaway, 1896), and toaster-stoves, "disc stoves" (hotplates), chafing dishes, coffee urns, and the like. A different class of appliances based upon the electric motor, included the electric fan (Wheeler, 1882), the washing machine (Hurley, 1907), and the vacuum cleaner (Spangler, 1907). By 1910 several thousand electrical power stations were in operation and one in ten American homes, primarily urban, had been wired for electricity.

In the class of communication products, Bell's telephone (1876), Edison's cylinder phonograph (1877), Berliner's disk phonograph (1888), and Jenkin's motion-picture machine (1894) initiated a revolution that would significantly transform American life. By 1910, there were more than 10,000 nickelodeons in operation in the United States. And when Marconi sent his first radio signals across wireless space (1895) it became evident that electricity could fuse the Americans into a homogeneous society.

In transportation, the horse and carriage and the steam locomotive were being challenged by electric trolleys (Van Depoele, 1884) and by electric (Morrison, 1892) and gasoline-powered automobiles (Duryea, 1892). Even the safety bicycle (Starley, 1884), which had been all the rage in the 1890s, was being displaced by the innovative powered vehicles. After the Wrights' first flight in 1903 it was predicted that flying would be "the only recognized means [of travel] within twenty-five years." (166)

Americans were particularly fascinated by the speed that technological transportation made possible. In 1900 an editor of the *New York Times* wrote that "experience with the bicycle [has] shown very clearly that the average human being is so constituted that he has an insatiable passion for high speed which makes it extremely distasteful to him to go slow when he can possibly go fast." (161) Another editor wrote that "it is characteristic of the American Nation that they

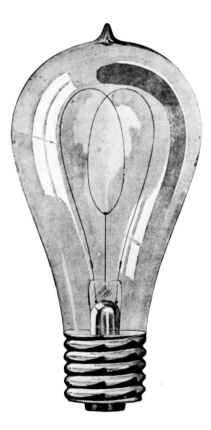

The Electricity Building
was one of the major at-
tractions of the 1901 Pan-
American Exposition in
Buffalo. Its Spanish architectural styles helped to
make representatives from
South American countries
feel at home and to con-
vince them to purchase
electrical products from
participating firms from
the United States. Elec-
trical World, March 24,
1900.

By 1900 the typeform
of the incandescent lamp
had been established.
Electrical World, De-
cember 8, 1900.

The ideal kitchen in 1908, if one could afford it, used electrical cooking appliances that occupied less space than the great coal or gas cooking stove. Most of the broilers, ovens, and saucepans that were available at first were simple geometric forms produced in small quantities by sheet-metal artisans. Architectural Record, *June 1908.*

After 1910, as electricity became commonly available in cities and towns, electrical devices for practically every domestic need appeared. These vernacular products, for the most part, display the unassuming geometry of limited production. Popular Mechanics, *1908.*

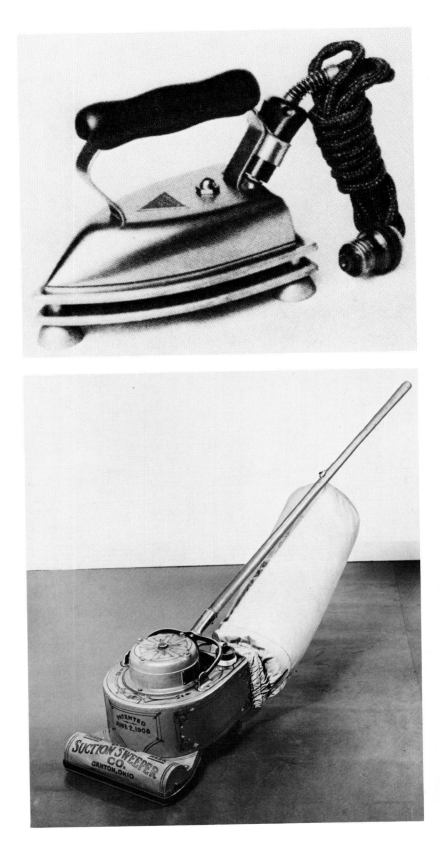

The invented form of the
electric iron was quickly
adopted as a typeform
and persisted for nearly
30 years. It combined the
cast sadiron plate with
a simple sheet-metal shell
and a turned wooden han-
dle. Its epitome was this
American Beauty, manu-
factured by the American
Electrical Heater Com-
pany of Detroit for some
30 years with very little
change. National House-
wares Manufacturers
Association, The House-
wares Story.

The Model O vacuum
cleaner was the first electrical machine manufac-
tured by the Hoover
Company after patent
rights were purchased
from Spangler. It was
essentially a "shop-form"
product, constructed
primarily by hand methods
in a sheet-metal shop, with
a brush roll and fan made
of die-cast aluminum
and an oil-treated sateen
bag. Hoover Company.

The first typeform for this
1910 desk telephone
appeared before 1900,
when a central operator
connected all calls. In-
structions to central were
transmitted electrically
by the hook. The manu-
facturing technology was
essentially lathe-based.
Reproduced with per-
mission of the American
Telephone and Telegraph
Corporation.

Thomas Edison's 1907 cylinder phonograph represents the invented form of the product. Despite its original popularity and romantic form, it was replaced by Emile Berliner's disk phonograph. Smithsonian Institution.

The disk phonograph, even in its invented form, quickly displaced the cylinder machine because disks could be reproduced more economically than cylinders and because it was easier to operate. In 1896, Elbert Johnson added a spring-wound motor. Smithsonian Institution.

The first patent for the typeform of the phonograph cabinet was issued to John C. English in 1910 for this comfortable blend of French cabriole legs and a quasi-mansard roof. U.S. Patent 41,008, November 29, 1910.

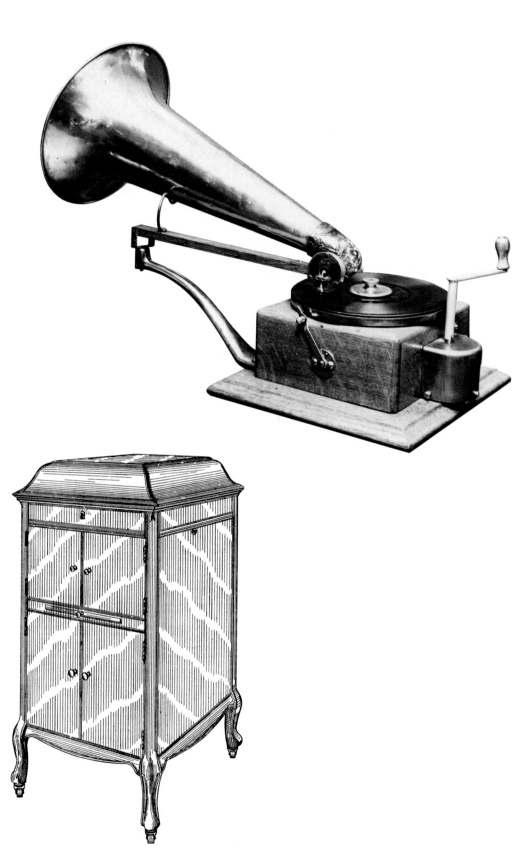

The first major exhibition of the American Automobile Club, held at New York's Madison Square Garden in 1900, was devoted almost entirely to electric vehicles. It was noted at the time that the newer designs were much more graceful, and that the chopped-off "where is the horse?" look was disappearing. Electrical World, November 10, 1900.

The first successful gasoline-engine vehicle was this single-cylinder horseless carriage designed by Charles E. Duryea and built by his brother J. F. Duryea in 1893. The third version of this design won the first American road contest for automobiles. *Library of Congress.*

The Electric Vehicle Company of Hartford, Connecticut, displayed this neat little Columbia runabout, capable of running 100 miles without a stop for fuel or water, at the Automobile Club's 1900 show. *Electrical World, November 17, 1900.*

FIG. 2.—BODY SHOP, NEW HAVEN FACTORY.

FIG. 6.—WHEEL ROOM, NEW HAVEN FACTORY.

FIG. 3.—ASSEMBLING ROOM A, MAIN WORKS, HARTFORD.

FIG. 7.—MAIN WORKS, HARTFORD.

FIG. 4.—ASSEMBLING ROOM, EAST WORKS, HARTFORD.

FIG. 8.—ASSEMBLING ROOM, MAIN WORKS, HARTFORD.

FIG. 5.—FORGE SHOP, NEW HAVEN FACTORY.

FIG. 9.—INTERIOR OF MAIN WORKS, POWER FORGE

The Electrical Vehicle Company of Hartford absorbed the Hartford Cycle Company as well as the New Haven Carriage Company, where a large proportion of its vehicles were manufactured. At this point, the methodology of manufacture was a long way from mass production. Electrical World, May 26, 1900.

The Evinrude company promoted its prototype gasoline-powered outboard motor in 1908 with such studio-posed photographs as this, which seems intended to demonstrate the ease with which women could handle such complicated equipment. Library of Congress.

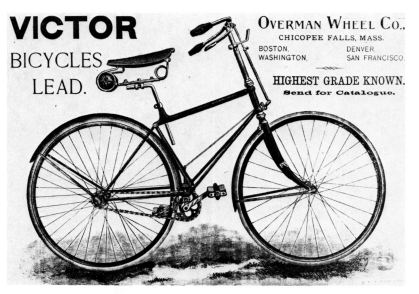

The safety bicycle, which displaced the high-wheeler, made personal transportation for men, women, and children safe, efficient, and enjoyable. This invented form established a typeform that persists after nearly a century as the simplest mechanical means of personal transportation. New York Public Library.

This romanticized illustration by Will Bradley, from Life magazine in 1896, is a tapestry of ornament in an Americanized interpretation of Art Nouveau. New York Public Library.

Orville Wright piloted the Flyer in its first successful flight at Kitty Hawk, North Carolina, in 1903. Smithsonian Institution.

The first practical airplane to be built, demonstrated, and reproduced by the Wright brothers for a client (the U.S. Signal Corps, in 1909). Smithsonian Institution.

believe in speed, and the faster business or pleasure vehicles are driven the better they like it, and it is this desire to 'get there' that causes the trouble." (164) Those who were more cautious proposed that regulations or regulating devices be created to fix maximum speeds "to prevent the growth of an evil which, when once grown, will defy suppression or regulation." (161) One critic was so brash as to suggest that every car should be registered, licensed, and fitted with a speed regulator. The *Times* opined that "The automobile is with us to stay . . . and since it is the special modern vice to try to get over ground with swiftness for the mere pleasure of the motion, the only thing to do is to make the instrument of our vice as little obnoxious to our fellow-man as possible." (165) J. B. Walker, editor of *Cosmopolitan* magazine, acknowledged the future of the automobile by noting that Americans were "on the verge of a revolution in our methods of living." Wrote Walker: "To be able to travel, with a fair regard for safety, along a smooth highway at a pace that rivals the speed of the railway accomodation train, if one is willing to take the risks or the police tolerate him, is now fully within the power of the owner of the newest type of automobile carriages. . . . a man may live twenty-five miles out of the city, and . . . reach his office by the most delightful means of locomotion within an hour." (160)

For a while, that the automobile worked at all and could be operated with reasonable reliability was sufficient. However, very soon the public began to insist that the gaunt mechanical contraption should show its speed and power in form as well as action. Some felt that "our automobiles have been truly monsters of ugliness" and suggested that any manufacturer who invested his product with "graceful curves and fine proportions," turning out "a beautiful as well as a speedy and comfortable machine," would be well rewarded. (162) One critic wrote that automobile builders had "not attended to the requirements of art, but turned out a monster which adds to its manifold injuries the result of ugliness." Another complained that the automibile was "still a raw invention . . . a vehicle which is still unadapted to all the purposes for which it is built." He felt that "few objects are more hideous than

the high four-wheeler, the 'buggy' that lacks a horse," saying that it was "like an old shoe," could "scarcely lay claims to beauty," and did not "suggest by its lines the idea of swiftness." (165) The term *streamlining,* coined in 1893 in connection with hydrodynamics, was beginning to be applied to automobiles. By 1909 it was being suggested that streamlining would help to conserve fuel as well as to carry off the noxious fumes of the gasoline engine. It was becoming evident that the automobile would have to abandon its dependence upon carriage-building methods and seek a new form and manufacturing methodology that were commensurate with public expectations.

Most manufactured products of the era had yet to seek, let alone find, unique typeforms. Automobiles were horseless carriages, cast-iron parlor and cooking stoves were baroque idols or altars, cash registers were Renaissance jewel safes, lamps were Art Nouveau gardens, and furniture was still lost in Victorian reverie. Many manufacturers were more preoccupied with sales figures than with aesthetic form and borrowed styles freely from any source that promised that the volume of products marketed would be in line with increasing production capacity. It was enough for them that by 1900 the United States had surpassed all other countries in the volume of manufactured products that were finding purchasers at home and abroad, not because they were handsome but because they were often innovative and always cheap. However, an undercurrent of dissatisfaction and even embarrassment was emerging about the lack of genuine aesthetic quality in American manufactures. In the Paris Exposition of 1900 the United States exhibited 7,000 items, more than any other country — most of them clever labor-saving devices and inventive mechanisms. "It is only when it came to the matter of adornment, whether for the person or the home," it was reported, "that there was doubt [of quality]." (163) Some critics suggested that the Europeans were really ahead of the Americans because they were learning to counteract the success of American ingenuity and low cost by improving their own products on the side of art so that the people would be willing to pay more for them. Americans must no longer "trust to the cheapness of machine-made goods," insisted a 1902 editorial in the *New York Times;* goods must be made "decidedly more attractive to the eye." (162) Thus, the concept of infusing manufactured objects with artistic qualities was publically proposed early in the century as a matter of sound business strategy.

A number of organizations, such as the Artist-Artisans Institute, the Decorative Art Society, the Society of Applied Arts, and the National Arts Club, were established to demonstrate that the same virtues that had brought American manufactures to the fore for their utilitarian attributes could be applied to improving their aesthetic qualities. In 1900 the National Board of Education reacted to their position by establishing a committee to develop a proposal for a more comprehensive program of American education in the industrial arts. This effort was motivated not so much by a desire to elevate public taste as by the belief that an increase in the artistic quality of manufactured products would be of economic benefit to their makers as well as to the national economy. Although the committee's three-volume report recommending that government and industry should support such design education was distributed widely, none of its proposals were put into effect. The general reaction was that, since the best design was thought to be foreign, there was little need for Americans to invest in the development of their own industrial-arts capability. Nevertheless, the growing importance of the designer in America was evident in the announcement by the United States Commissioner of Labor in 1904 that a stove manufacturer was so concerned about the aesthetic qualities of his product that he paid a well-known designer $5,000 to design a kitchen stove that would "not offend the eye of the day laborer." (211)

It was natural, perhaps, to expect that those manufactured products that had distinct social and cultural precedents would rely heavily upon historical styles and would borrow forms and details (such as cabriole legs, swan's-neck spouts, pineapple finials, and rococo handles) that could be worked into products that carried

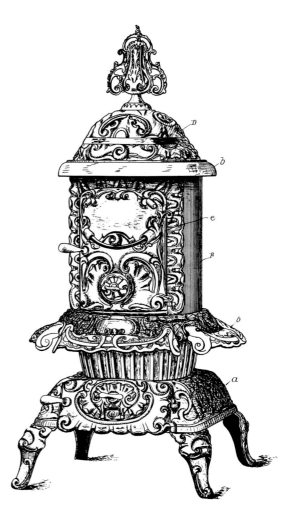

The parlor stove, as exemplified by this patent drawing by George Cope and William Bertram, became the focus of the home. This design was assigned to the Kalamazoo Stove Company. U.S. Patent 36, 312, May 5, 1903.

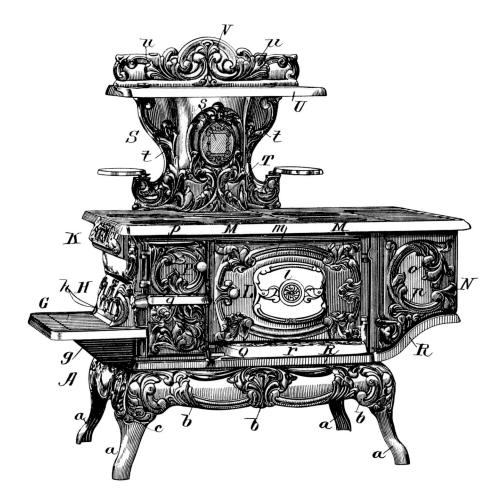

Thomas Kennedy and
Edward King of Rochester,
New York, were the most
sought-after designers
of cast-iron products.
Their work shows a rhyth-
mic consistency that is
suited to the structural
requirements of cast-iron
stoves. U.S. Patent
35,932, June 3, 1902.

The invented form of
the cash register (by
E. Ringold) looked like
a wooden clock with keys.
By 1900, however, it had
evolved into this treasure-
box typeform, decorated
with cast bronze plates.
U.S. Patent 36,831, March
1, 1904.

Victorian furniture seemed
to be more for show than
for use. The pieces were
often displayed in drawing
rooms or parlors, which
were opened only for for-
mal and ceremonial af-
fairs. Harper's Monthly,
July 1876.

The electric grill featured
in this 1908 General Electric advertisement had yet
to achieve the traditional
styling of the electrified
kettle, urn, and chafing
dish in the background.
Library of Congress.

the exclusive flavor if not the expensive substance of the originals. This practice established a pattern that still prevails. Manufacturers of such products as chinaware, glassware, flatware, holloware, textiles, furniture, and lamps continue to borrow traditional forms or foreign fashions in order to provide an array of seductive styles for the upward-striving consumer. One of the most successful examples in this area is the Gorham Company's "Chantilly" flatware pattern, introduced in 1896—a skillful blend of rococo and Art Nouveau forms designed by the company's staff with help from the English émigré William C. Codman. The pattern is still being manufactured and sold by Gorham and, as the most popular pattern of silverware in the United States, is echoed in lines offered by many other manufacturers.

Another category of products being manufactured at the time included technology-based appliances, such as electric fans, irons, toasters, and vacuum cleaners, that had no historical precedents but were becoming essential to daily life. It never occurred to their makers that the unique services these things provided deserved original typeforms. Rather, manufacturers sought to mask their identity by adopting traditional forms and details. Despite their artificial elegance, however, early appliances were clearly "uncomfortable" in their ill-fitting and often inappropriate costumes. Beneath the aristocratic presumption one could sense the basic humility of the new utilitarian forms that would one day achieve their own identity. It would seem that, in the process of being transformed into manufactured products for public consumption, most inventions go through a period of historical pretense before their own character becomes strong enough to slough off their arbitrary camouflage. In time, however, utilitarian products inevitably acquire an aesthetic of their own as familiarity transforms them into unique forms that contain the essence of their function. Some critics of the early twentieth century lent weight to this view by warning that when the machine would take over, those who made artistic products would have to either join industry or look elsewhere, because the drive to capture larger markets would inevitably lead to ill-conceived baroque products as increasing competition would require novelty. In an interesting analogy of the time, machine-made art was compared to a political machine that brings out the vote and expresses the will of the people but in the end tends to kill statesmanship. Once the Arts and Crafts movement failed to reverse the drive toward machine production of the amenities of human comfort and convenience, its influence was dissipated in two directions. On one hand, it broadened its interest in the machine as a convenient means of replicating traditional styles that could provide a sort of synthetic opulence within the economic reach of everyone—the Industrial Arts style. On the other hand, it deepened its search for a new aesthetic that was in harmony with technology—the Machine Art style.

If most Americans were too close to their machines and the products of their factories to recognize that there was a unique cultural quality inherent in machine-made forms, a few perceptive architects were not. Frank Lloyd Wright was convinced that the machine was here to stay because it had become indispensible to the economy. The Austrian Adolf Loos (1870–1933), in contradiction to the Viennese fashion for the neo-romantic as a subjective substitute for historicism, spoke out for the basic democracy of manufactured products. Having spent most of the 1890s in the United States and England, Loos returned to Vienna convinced that a new spirit of form had come into being by way of manufacturing processes, avowing that cultural revolution was equivalent to the removal of ornament from articles of daily use. Loos was also one of the first to realize that there was a basic difference between art and design. His experiences in the more industrialized countries removed all his "prejudices against the products of [his] own time" and convinced him that, whereas art was self-serving and therefore likely to be degraded by design, the beauty of designed objects was vested in their appropriateness to manufacture and use. Loos declared that it was a degradation and a prostitution of art to mix it with design, and that it was barbaric to waste art on an article of utility. There was no absolute beauty in a utilitarian object, he believed; its beauty lay solely in its appropriateness.

The "Chantilly" pattern
of sterling flatware was ini-
tially produced by the Gorham company in 1895.
Skillfully combining ro-
coco and Art Nouveau ele-
ments, it was conceived
as a generic effort by the
company's design staff
with the assistance of
William Codman. Courtesy
of Gorham/Textron.

The "Chantilly" pattern,
produced today by the
Gorham Division of the
Textron Corporation, is still
the best-known and has
been the best-selling sterling flatware design in
the world. Courtesy of
Gorham/Textron.

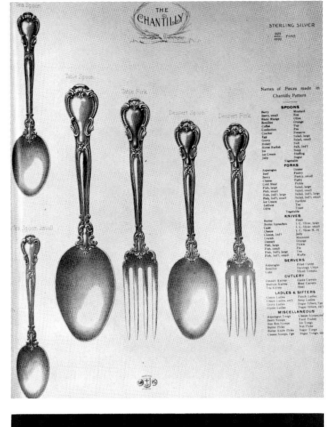

The Germans, who had followed the Arts and Crafts movement and industrial developments in England and America closely over the years, sensed that a new philosophy of form was developing that promised to bring the qualities of craftsmanship and the potential of manufacturing closer to one another. In 1896, the Prussian Board of Trade dispatched Herman Muthesius (1861–1927) to London to report on the changing design and architecture scene with an eye toward improving the international trading position of Germany. Muthesius returned in 1903 impressed by the movement toward a more direct and honest use of materials and industrial processes and the ingenious naivete by which the English and Americans were forging ahead in the manufacture of vernacular products. Some German designers, like Richard Riemerschmid, considered the machine to be little more than a source for style, but Muthesius pressed his government to take a more objective approach. In response to his urgings the Prussian government established in 1907 the Deutscher Werkbund, which brought together artists, craftsmen, patrons, and others in an attempt to improve production by fostering cooperation between art, industry, and the crafts.

Muthesius, as head of the Prussian Board of Trade for Schools of Arts and Crafts, was able to make certain that those architects whose works reflected his own design sympathies were appointed to head the most important schools. Bruno Paul (1879–?) was named to head the Industrial Art School in Berlin, and Peter Behrens (1868–1940) the one in Düsseldorf. The year 1907, when Paul was commissioned by the Dresden cabinetmaker Karl Schmidt (?–1948) to design a line of low-cost furniture for machine production and Behrens was hired by the AEG electrical company to direct the quality and character of its buildings, graphics, and products, is taken by some as the beginning of industrial design. Behrens's office provided practical training for Walter Gropius (1883–1969), for Ludwig Mies van der Rohe (1886–1969), and for Le Corbusier (1887–1965), who spent several months with them.

The Deutscher Werkbund provided a valuable arena that sharpened the distinction between the artist and the designer and clarified their roles in industry. The Belgian Henry Van de Velde (1863–1957), then director of the Weimar School of Arts and Crafts, believed that the role of the artist in industry must not inhibit his creative freedom; the position of the designer was championed by Muthesius, who insisted that the demands of machine production establish the rules that direct design.

The activities of the Werkbund were not unknown in the United States. In 1912, as part of its program to improve the taste of the public as well as that of manufacturers, the Werkbund sent an exhibition of the work of its members to the Newark Museum of Art and subsequently to other American institutions. John Cotton Dana, then director of the Newark Museum, who had invited the show because of his strong interest in vernacular design, believed that "beauty has no relation to price, rarity or age"—all values normally associated with aristocratic products. He was aware that the Werkbund's interest in exhibiting in the United States was centered upon the potential market for German products, but he hoped that the exhibit would also serve to awaken American manufacturers to their own design responsibilities.

It is convenient to assume that modern industrial design began with the recognition by some individual that machines as well as machine-made products could evince a unique machine-art aesthetic. However, it is more reasonable to acknowledge that industries in the United States and in other countries had been producing vernacular products that were devoid of applied style for many years before such utilitarian forms were discovered by artists and architects and elevated to the status of machine art. The unique characteristics of machine art included minimalism as a commitment to economy of means and materials, geometricity as obeisance to technological rationalism, and anonymity as a denial of human expression. Its creed was technical perfection of form and mechanical performance. The essential principle of functionalism (the subservience of the product to need) had been recognized by

Tocqueville and confirmed by Greenough in the nineteenth century, but it remained in the background until well into the twentieth century, when the arts of manufacture were accepted as distinct from the industrial arts.

That the 1900 study of the need for industrial-arts education in the United States did not result in a program of action did not mean that the demand was not justified. An overwhelming sentiment continued to be expressed in newspapers, magazines, and journals in support of broadening the original call for industrial-arts education to include what were then called vocational studies. This time, the pressure was intensified by the arrival of a great wave of immigrants (nearly 8 million between 1900 and 1910), most of whom were from countries that did not have established industrial systems. These people could not be put to work in the industrial-arts industries (those that manufactured furnishings and decorative products) unless their education was subsidized by government in some way, with special emphasis upon the particular needs of local industries.

The first real break for vocational education came in 1905 when the governor of Massachusetts established a committee to study the problem. The committee found that there was a lack of skilled artisans in industry and proposed that the state should bear all or part of the cost for such education. It further recommended that public schools should add practical education in industrial, mechanical, agricultural, and domestic arts by modifying their programs to include elements of production in industry and agriculture as well as mathematics, science, and drawing. In short order, other states such as Vermont, Indiana, New Jersey, Maine, Michigan, and Wisconsin also joined the movement by supporting legislation in favor of federal and state-supported vocational education.

In 1909 the National Society for the Promotion of Industrial Education was founded at the Cooper Union in New York, with Henry S. Pritchett, then president of the Massachusetts Institute of Technology, as its first president. Its greatest successes were in reconciling the differences of opinion between manufacturers and labor leaders on one hand and others promoting education for the agricultural and domestic arts and in laying the groundwork for eventual congressional approval of federal support for vocational education. They were joined by others who had a vested interest in the industrial-arts industries. In 1914, the Grand Rapids furniture industries undertook their own study of what had to be done to develop native industrial artists. Their report, which was distributed to members of Congress and others on behalf of the National Board of Education, compiled statistics on every important industrial-arts area and outlined for the first time the procedures by which industrial artists could be employed by industry to design products that would be pleasing and acceptable to the public. Although the report undoubtedly added to the growing pressure on Congress to support vocational education, somehow its special interest in industrial-arts education was lost in the move toward technical rather than aesthetic training.

After the failure of one attempt, Congress approved a committee to establish a permanent commission on federal aid for vocational education. This time Congress was astute enough to recognize the increased political leverage that the program would have if it were broadened to include the special interests of the agricultural and domestic arts, and therefore included representatives from southern states. With Georgia's Senators Hoke Smith and D. M. Hughes as leaders, the commission studied conditions and needs in the United States and the existing vocational education programs abroad. Its recommendations were that federal grants should be matched with state grants to train, and to pay part of the salaries for, vocational-education teachers in the areas of agriculture, industry, and home economics. This support was to be directed at students of less than college grade between the ages of 14 and 18 who were enrolled in programs that combined theory and practice on an equal basis. President Woodrow Wilson signed the Smith-Hughes Act (as it came to be known) into law in 1917 and established the Federal Board of Vocational Education to administer it.

Mass Production and
Concern for Design

The Smith-Hughes Act has remained essentially unchanged since then. With the exception of a few isolated cases, it does not offer any useful support to higher education for the education of designers for industry. Its focus remains primarily on the secondary-school level, where it serves trade, agriculture, and domestic science.

To this day children in American schools receive very little formal education in the aesthetic and humanistic quality of the products of technology. Although they will live out their lives in a technology-based environment, they are essentially ignorant of its cultural and functional values. It is no wonder, then, that as mature citizens, they should continue to be abused by those industrial-art industries and manufactures that are equally insensitive to indigenous expression and, therefore, content to emulate the culture of other times and places.

I have striven toward manufacturing with a minimum of waste, both of materials and of human effort, and then toward distribution at a minimum of profit, depending for total profit upon the volume of distribution. . . . (33, 19) A man ought to be able to live on a scale commensurate with the service that he renders. (33, 11)

Henry Ford, 1915

There is a natural relationship in manufactured products between the properties of the materials that are used, the machines and processes that will be employed, and the form of the product that will result. Moreover, each of these elements must be in a constant state of change if it is to maintain dynamic equilibrium with the others. At the same time, each is not only struggling for survival against variants of its own kind but is also under challenge from other areas that seek to displace it. In a technological society that is open to change, there is, therefore, an inescapable pressure upon the manufacturer to reduce the quantity and quality of the materials and the human and synthetic energy that are being consumed without reducing the value of the product that results—in other words, to do more with less.

The net result of such competition is that, as the manufacturer's margin of profit begins to shrink, he is obliged to expand the volume of production in compensation. In order to be sure of a market for this increase he must appeal to a broader buying public, not only in terms of price but also in equal value and improved beauty. The concept of modern mass production and industrial design begins at this point.

The practice of industrial design must anticipate every eventuality in the development of products or product systems that can be manufactured and distributed economically in order to meet the physical needs as well as the psychological desires of human beings. Rather than being tied to the one-on-one relationship that is characteristic of traditional handicrafts, industrial design is based on the concept of one on many—that is, one person or team of persons takes on the challenge of conceiving, developing, and manufacturing a product that will be acceptable to a great number of people, usually unknown to the makers. The investment cost of starting up a manufacturing enterprise and sustaining it to final production and sale is so great that those who are willing to take the risk must take every precaution that their product will be as economical to purchase and as pleasing to behold as it is efficient, convenient, and safe to use.

The first dramatic signs of mass production began to appear before World War I when it became evident that automobiles were being designed for more modern production methods. Besides reducing the time needed for manufacture, these new methods opened the way for new forms. In 1914, at the fourteenth annual automobile show at the Grand Central Palace in New York, it was acknowledged that "streamlines in their fullness are here." (169) Before this time, automobile bodies had been hand-fabricated by carriage-making practices or, where sheet metal was used, by shaping each piece with traditional sheet-metal tools that were only capable of rolling, folding, seaming, and beading planar surfaces. Crowned surfaces could only be obtained by laborious hand pounding and fitting to a wooden form. However, the development of heavy presses that could stretch flat steel into three-dimensional shapes made the streamlined form possible. "Long sweeping lines," it was stated at the time, "are pressed into the steel under great pressure, and there is no hammering out at any point in order to develop the curves. Consequently, the steel is of uniform thickness and strength at all points. All joints are electrically welded at the doors so that the finished body is actually composed of a single piece of sheet steel without seams." (171)

In 1908 Ford began manufacture of the Model T on the premise that the automobile of the future should be affordable to the masses. The first Model T sold for $850—the same price as the two-passenger, two-cylinder, four-horsepower Fordmobile of 1903. By 1915, when the Ford assembly line was shown to the public for the first time at the Panama-Pacific Exposition in San Francisco, Ford had shifted his manufacturing methods to automatic machines and moving production lines. He was able to demonstrate convincingly that his new methods would bring prices down to a level that would put the automobile within the reach of everyone.

The principle of an automobile assembled from components manufactured elsewhere, which lies at the heart of mass production, had been pioneered by Ransom E. Olds as early as 1899. However, the process was refined in 1913–1914 at the Ford Motor plant at Highland Park, Michigan, as a twentieth-century extension of the well-established practices of standardization of parts and specialization of labor. Dramatic proof of its effectiveness was the reduction in the final assembly time of the Model T from 12½ to 1½ hours per car. The new principles, as outlined by Bruce Nevin, included the following:

- There was planned, orderly, continuous progression of the product being manufactured through the factory. (Automobile production lines still move along the assembly line at a speed of something like three miles an hour, with every aspect from mining the ore to consumption of the final product geared to this inexorable speed.)

- The mechanics remained at their place of work as the product moved past them. They were trained in the task to be performed and provided with specialized tools. Accessory parts that they would need were delivered directly to their workplace.

- All of the operations were analyzed so that the mechanics could do their work with a minimum of fatigue within the shortest period of time. The parts that they were responsible for were designed, with every possible deviation recognized and neutralized, so that they could be efficiently and accurately installed.

- The mass-production system was designed to operate 24 hours a day in order to make full use of the manufacturing facility.

Ford's mass-production methods guaranteed the then-unheard-of wage of $5 for an eight-hour day, instead of less than that for the customary nine-hour day (this enabled Ford to run three shifts of workers a day instead of two, thus increasing the productive capacity of his plant by nearly 50 percent). This wage policy incited riots among workers for whom Ford had no jobs to offer as well as workers at other factories. It raised a

This replica of the brick
shed in Detroit in which
Henry Ford built his first
automobile stands in
Greenfield Village at Dearborn, Michigan. Courtesy
of Ford Archives, Henry
Ford Museum, Dearborn,
Michigan.

Henry Ford success-
fully operated his first
automobile—the Quadri-
cycle, a two-cylinder,
four-horsepower vehicle
for two persons—on the
streets of Detroit in 1896.
Courtesy of Ford Archives,
Henry Ford Museum,
Dearborn, Michigan.

The 1913 brass-radiator
Model T was the epitome
of the "Tin Lizzie." By 1913
it was being stripped
down and analyzed for
mass production. Cour-
tesy of Ford Archives,
Henry Ford Museum,
Dearborn, Michigan.

The final assembly line
for the Model T in 1914
at Highland Park brought
the assembled body down
a ramp to be joined to the
chassis. This was the final
stage after the many sub-
assembly lines inside the
plant had produced the
engine, the gas tank, the
dash-board, and other
components. Courtesy of
Ford Archives, Henry Ford
Museum, Dearborn,
Michigan.

storm of protest from other manufacturers, who complained that it was utopian and contrary to experience. However, Ford stood firmly on the position that by paying his workers higher wages for fewer hours of work he was making better use of his facilities and was bringing the product within the reach of his own workers. (Its original selling price was less than $1,000, but with mass-production methods the price of a Model T had dropped to less than $300 by 1914.) Ford rejected any criticism that he was exploiting his workers. Rather, he believed that he had democraticized industry, and warned that unless industry managed to keep wages high and prices low and provide its workers with shorter hours it would limit the number of possible customers. It was obvious to Ford that if workers were given more disposable income and more free time, they would become more mobile, and that they would aspire to own their own homes and seek to outfit them with all of those manufactured conveniences that support a higher standard of living.

Thomas A. Edison, Henry Ford's lifelong friend, praised the manufacturing philosophy of the Ford Company:

When we use machines instead of humans and have a single apparatus to do the work of 250 men, then employees will enjoy real benefits. . . . The time is passing when human beings will be used as motors. We are today putting brains into machinery, and are replacing by machinery the energies of thousands of humans with only a few men to see to it that the apparatus keeps working. . . . Improved machinery . . . would necessitate men working fewer hours, and at the same time would enable them to accomplish much more. . . . Machinery is the salvation of the American manufacturer. (170)

Although Henry Ford has been criticized for a lack of concern with aesthetic values, he was acutely aware of the impact his automobile had upon the public and the affection with which they tolerated its idiosyncrasies. By 1915 more than half a million "Tin Lizzies" had been built and sold, and the gaunt, look-alike car had entered American folklore. Over the years of its production, from 1908 to 1926 (when a more stylish Chevrolet forced Ford to abandon its manufacture), the Model T gradually lost its awkwardness as it evolved into the type form that has made it

a treasured artifact of the era. More than 15 million Model T's were sold over the years—a record surpassed only recently, by the Volkswagen "Beetle."

It was inevitable that other industries would adopt the Ford Motor Company's manufacturing methods and that, as the volume of products increased proportionately, competition would oblige them to take consumers' taste preferences into account. At first, manufacturers designed a product to suit their own tastes or assigned the responsibility to employees who showed some sympathy for style or aesthetic value or to modelmakers and pattern carvers who were familiar with prevailing fashions. In time, however, there emerged independent designers, usually specializing in one product area, who sold their services alternately to competing manufacturers. Prior to this period, most design patents had been in the areas of the decorative and industrial arts. Now more and more were being issued for the casings or forms of manufactured products—what may be classed as industrial design.

The manufacturing establishment was also becoming aware of the value of corporate identities and trademarks in establishing and maintaining public allegiance. Henry Ford conscientiously renewed the script form of his name that had been in use since 1895. The famous mark of the Victor Talking Machine Company was designed and filed for protection by Eldridge Johnson, who had purchased the rights to manufacture what he called the Gramo-o-phone from Emile Berliner, before the company itself was named. The Coca-Cola Company, already in possession of a distinctive trademark conceived by Frank Mason, redesigned its bottle so that it too would serve as a point of public recognition. And Charles Eastman, already in possession of the coined word Kodak for his company, named the $1 "Brownie" camera after Palmer Cox's busy little elves, who may be taken as cartoon ancestors to Walt Disney's Mickey Mouse and Charles Schulz's Charlie Brown and Snoopy, who are busily peddling manufactured products to children today.

"I wish we had one of those."

1916

Henry Ford's trademark, first used in 1895, was in the bold Spencerian script that was popular at the time. It was distinctive enough to be admitted to the principle trademark register of the patent office, as well as to ward off imitators. Ford Motor Company.

The famous Victor trademark showing a fox terrier listening intently to "His Master's Voice" was designed by Francis Barraud, an English artist. The original painting depicted Edison's cylinder machine. After the Edison company turned it down, Barraud sold the idea in 1899 to the Gram-a-phone company and painted in its Berliner disk machine. U.S. Patent 84,890, May 26, 1900.

Although the Victor trademark was instantly popular, the company acquired other versions, such as this one, in order to cover what it considered to be a reasonable range of alternatives. U.S. Patent, 40,224, March 19, 1903.

A second trademark filed by the Victor Company. The application made a careful point about the importance of the rose. U.S. Patent 40,225, March 19, 1903.

The distinctive "hobble-
skirt" Coca-Cola bottle,
introduced in 1916.
Author's collection.

The Eastman Kodak
Brownie camera capital-
ized on the popularity
among children of Palmer
Cox's amusing elves as
well as suggesting the
name of its designer,
Frank Brownell. Cour-
tesy of Eastman Kodak
Company.

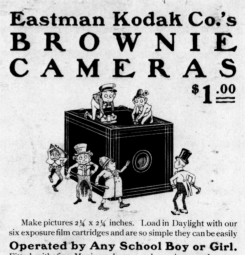

It is not an exaggeration to point out that the main body of mechanical and electrical devices, appliances, and machines that form the backbone of American technological convenience came into being between 1875 and 1915. Their shapes were explored and revised in a vigorous variety of designs until they achieved more stable typeforms. The period since then has been, except for isolated examples, essentially one of consolidation and refinement to meet changing human requirements or to take advantage of new materials and manufacturing technologies.

The manufacturers of utilitarian products were plagued by the same problem of design piracy that afflicted the decorative and industrial arts. Some felt that originality in design did not pay because it was too difficult to get adequate protection for an innovative concept. The marketing advantage seemed to lie with those who borrowed successful ideas without risking investment in untested ones. This disturbed both American and foreign manufacturers. The U.S. State Department was being bombarded with complaints from other countries that the designs of their manufacturers were being pirated wholesale, and American manufacturers were counterclaiming that their designs were being stolen by companies that collected the best in American design and reproduced them for sale in Europe. For the United States, loss of honor and international humiliation were implied in the announcement that some European manufacturers would not exhibit in the Panama-Pacific Exposition in 1915 for fear that their best ideas would be pirated by the Americans. As a result, a general demand developed for better design patents or for regulations that would give "exclusiveness of manufacture and sale to the originator of a design in dress fabrics or accessories, decorative fabrics, the distinguishing form of a machine, or any form which constitutes merchandise value." (168) In 1913 the Commissioner of Patents, Edward B. Moore, came out publicly in favor of a modification of existing regulations to protect property in "industrial design." (167) (This is the earliest known public use of the term *industrial design* as a generic description referring to the distinguishing form of products that have marketable value. It appears that the term was used

nonspecifically for many years before it was preempted as a verbal banner for the new profession.)

In 1914 a bill for protection against design piracy was introduced in Congress by Representative Oldfield of Arkansas. This bill, which was supported by the National Registration League and the Federation of Trade Progress of the United States, recognized that under existing conditions decorative design elements were specified as being embodied in a particular product and that existing laws did not provide protection when they were applied to other products. It promised to provide a truly flexible system of design protection that would prohibit such activities. In the testimony that was given during the hearings, a Kentucky manufacturer of cooking stoves complained that his products were being duplicated and sold by nine southern factories. In rebuttal, stove repairmen maintained that the manufacturer only wanted to maintain a monopoly on parts for replacement and repair. The Gorham Company volunteered the testimony that one of its own products, a silver almond dish that was originally sold at $13.50, had been copied in a lighter gauge of metal for sale at $10.50, thereby reducing annual sales of the original from 12,000 to 8,000 units. The primary issues in the testimony were claims that design piracy tended to cheapen goods brought onto the market, that it penalized designers, and that it destroyed the will of manufacturers to invest in or produce their own designs.

Nevertheless, many manufacturers and merchandisers continued to insist that they had not only the right but also an obligation to their clients to copy European design ideas. To justify their position they pointed out that the complacent American art and design community had been taken by surprise by the vitality of the impertinent art forms of the 1913 Armory Show at New York. They insisted that the tide of expression continued to flow westward across the Atlantic, bearing images not only from the familiar past but also from the challenging future. As the United States began to prepare psychologically to enter World War I, a general concern was expressed that the war would interrupt the flow of taste and talent from abroad. Manufacturers and merchandisers resented any disruption of the source of their ideas. An editorial in the *New York Times* acknowledged the following: "For over fifty years these trades have obtained their talent from abroad. Foreign designers by the hundreds have sought employment and have found it in our factories. . . . While it could secure designers from abroad, industry was indifferent to the development of the powers of our native artists." The art trades themselves had "with scant patriotism lauded 'foreign designs' and emphasized the fact that their patterns were 'imported.' " (176) It was considered smart business to let other countries train designers and then to import them to serve American industry. Many American industries still hold to this philosophy.

The situation regarding industrial design in the United States is improving, but so far as we can estimate . . . we shall need after the war about fifty thousand more industrial designers than are now available or in training, and probably few can be imported. Each country will need its own. . . . We shall have to depend upon ourselves more than in the past, not only for designers but also for styles of design. . . .

Walter Sargent, 1918 (196, 442)

President Woodrow Wilson expressed the opinion of most Americans on the eve of World War I that the war (without American involvement) would benefit American agriculture and industry, and pragmatists like John N. Willys, president of the Willys-Overland automobile company, suggested that although they did not believe in capitalizing on another nation's misfortune it was evident that the closing of many European factories would leave the world's markets open to American manufacturers. Thus, during the three years that the United States was able to stay out of action in the front lines of the European war, it became an important source of supplies and machines for the countries that eventually became its military allies.

World War I gave the world its first glimpse of the awesome capabilities of mass-production methods as Henry Ford's production lines became the prototype for other factories producing war materiel. In building the highly mobile and militarily effective French 75-millimeter cannons the Americans refined manufacturing techniques and improved tolerances and production controls to within an accuracy of 0.002 inch—a remarkable achievement for the day. The production rate of machine guns climbed from 20,000 to 225,000 units a year, and once the production lines got rolling they were producing 500,000 rifles and over a billion rounds of ammunition a year. The American trucks that rolled out of the factories in great number were considered to be more suitable for military service than European ones because they had solid rubber tires, wider tread, and higher ground clearance.

American two-ton trucks were welcomed by the Allies in the first world war for convoy purposes. The demands of the front forced rapid improvements in these and many other manufactured products that would influence postwar production. Automobile Manufacturers Association.

The "America" was the first tank to be built in the United States. J. M. Miller and M. S. Canfield, The People's War Book.

Military and industrial success had a profound effect on the Americans' sense of power and prestige. It not only convinced them of their self-sufficiency but also imbued them with an aspiration to hold onto the world leadership they felt they had earned. The United States had made the world safe for democracy, it was claimed. Thus, for some Americans who were far from war's trauma of death and destruction, it seemed to be not only a profitable venture but also a glamorous adventure in a good cause. Irving Berlin caught their missionary fervor with "Over There."

The first world war also attracted attention to products other than military hardware. Before 1915, only dudes smoked "tailor-made" cigarettes; men smoked cigars and women did not smoke. Bracelet-watches, as they were called at the time, were not worn by men lest they be considered effeminate. The manufactured smoke offered the quick solace of tobacco to men in action, and the bracelet-watch alerted men in the first gleam of dawn as they prepared to go "over the top." By the end of the war it was apparent that mass production, which gave men comfort and victory on the front, would also provide for their needs and desires back home.

The first world war was a watershed in human relations. The war began in the gentlemanly style of the last century, with aristocratic pageantry and resplendent uniforms. It ended in twentieth-century fashion as men were driven into the mud of the trenches by machines. Tanks struggled on the ground while flying machines spied from the air and engaged in dogfights. Zeppelins introduced new atrocities when they dropped aerial bombs on the defenseless city of Antwerp. Hapless footsoldiers were raked by machine guns, incapacitated by gas they could not see, or decimated by shells from giant cannons dispatched from distances too far for the sound of their firing to be heard. Military technology had introduced the frightful concept of depersonalized combat. From now on, men could destroy one another without the visual horror of hate and hand-to-hand combat. With machines as the intermediary, men could now kill or be killed without the pain of conscience.

Thus, the war brought to the surface two factors that would influence the patterns of working and living in the following decades. One was the depersonalization of human relations as machines and the products of machines increasingly became technological intermediaries between people; the second was the belief that mass production would ensure success for everyone.

Unlike the United States, the combatant countries of Europe had recognized for years the financial importance of design in industrial-arts products manufactured for domestic and export markets and had supported national institutions to train artisans and designers. Even in the midst of war each was aware of and preparing for the economic battle for world markets that would certainly follow the cessation of military hostilities. Moreover, it was obvious that every European country would look to its own needs after the war before permitting the emigration of native talent to other countries. For example, at the onset of the war the Germans staged the Cologne Werkbund Exhibition to demonstrate to their own citizens and to foreigners their ability to manufacture products in the modern spirit with a neutrality of design that made them suitable for international consumption as well as mass production. However, during the war, almost an entire generation of German creative energy was lost, and with defeat imminent the country was obliged to take steps to ensure that artistic values would be given proper weight in all products to be exported. In order to preserve the talent that had been spared, the government ruled that no technical workers would be permitted to leave the country after the war. And, in order to reestablish the development of talent, the Deutscher Werkbund was authorized to lay plans for the reopening of its national art and design schools as soon as possible and to establish new schools. (The most renowned of these was to be the Staatliches Bauhaus, which would open in Weimar in 1919 under the direction of Walter Gropius.)

Nor did Great Britain permit World War I to interfere with her interest in and commitment to the industrial arts. For example, the English held an exhibition of their products in Paris in 1914 at the

Pavillon de Marson in order to demonstrate to the French that they had developed an efficient system of training industrial artists that would take the place of apprenticeship. In 1915, the newly organized Design and Industries Association convinced the British Board of Trade to sponsor an exhibition of the best products of its enemies Germany and Austria to be found in England, in order to encourage British industries to develop substitutes for products whose importation had been halted by the war. And in 1916, William Lethaby, who had been appointed as the first professor of design at the Royal College of Art in 1900, talked the International Studio in London into putting on an exhibition to demonstrate how a struggle for industrial supremacy would follow the war. The Design and Industries Association, which considered it a patriotic duty to keep British industry aware of the importance of design to the country, declared itself responsible for "the decency, economy and ingenuity of every object in our houses" and promoted the value of machine-made products that were aesthetically sound and demonstrated a fundamental fitness for use. (145)

The French had been sensitive to the importance of art values for many years, having established the Union Centrale early in the nineteenth century to bring the industrial and decorative arts together. Despite the severity of the war, France could not afford to let it interfere with her long-range goal of superiority in industrial-arts products. Within three months after the war began, France established a program whereby the children of an artisan who fell in battle would be educated in their father's profession at the expense of the state.

In 1917 the Comité Central Technique des Arts Appliqués conducted a conference at Paris to stimulate art industries. The report of the conference published in the *Arts Françaises* reported calmly (considering the times): "The basis of teaching in any craft or industry is the technique, of which the design and modeling are the chief modes of expression. . . . The objects are given forms always conditioned by the needs to which these objects must respond, and by the choice and requirements of the material employed in

their making. The decoration must always intervene only to complete the perfection of the craftsmanship and not to mask deficiency. The decoration comes in to define the significance of the forms and to accentuate the function." (172)

Conferences in other French cities also supported the need and importance of design education. The Marseilles Committee urged that prejudice against the education of artists be combated. In Dijon it was recommended that the class prejudice that separated makers from consumers be eliminated. In Toulouse a committee asked that the prejudice of war be put away in an acknowledgment of Germany's extraordinary industrial ability and that design continue to be taught and directed toward "utilitarian ends." Two months before the armistice, the French held an exhibition of design to honor the four brother designers Peignot who had died in combat. After the armistice, in an effort to regain her worldwide cultural leadership, France would again put into motion her plans, originally proposed in 1912, for a great international exposition of the decorative arts. The landmark exhibition, which would finally be held in 1925, was to cap the Decorative Arts style and introduce Art Moderne.

Prior to the war, the needs of American merchandisers and manufacturers for design in the industrial arts had been satisfied for the most part by the importation of products to be sold or copied and by the steady flow of skilled immigrants. As the war went on, Americans were chafing because their own sources of products and design had been interrupted and even halted. As a result, American sentiment began to run again toward the need to develop native design talent. In 1917, after a conference of 300 educators, designers, and manufacturers had expressed its concern about the quality of industrial arts in America, President Woodrow Wilson urged educators to keep in mind the postwar needs of the country by paying special attention to industrial-arts education. The Art-in-Trades Club of New York City responded promptly by establishing a series of prizes for design to be offered in public schools. Even patriotism was invoked as a proper justification for attention to design, with the suggestion that when the American soldiers came home from Europe they were entitled to find a new atmosphere in their homes. It was also proposed that it was necessary and patriotic to replace manufacturers' past practices of purchasing designs abroad each year. "We believe," one advocate wrote, "we have sufficient talent in America to supply the needs of our manufacturers. . . . The public should be interested in any organized effort to improve the character of industrial design in this country." (173)

The 1918 annual conference of the American Federation of Art (AFA), held in Detroit, was particularly concerned with the importance of the industrial arts to the United States. Keynote speaker Richard F. Bach, then curator of industrial art at the Metropolitan Museum of Art in New York, set the tone for the conference. "It is our patriotic duty to establish [industrial arts] schools during the war," he declared. "It will be an evil day for manufacturers and dealers after the war if American taste must again go to Europe for its industrial art products." Bach insisted that the industrial arts must be mobilized to achieve cooperation between manufacturers, dealers, designers, and the public in order to establish a higher standard of design in machine-made products. "We have gone into the European war for democracy," he said, "while at home we have allowed our development in the industrial arts to be ruthlessly autocratic. . . . If all products were hand-made few of us could afford them. Therefore, it is left for us to give the machine its proper place." (174) In response to the familiar argument that neither the public nor the manufacturers would be willing to bear the expense of training designers Bach replied, "I can say without scruple that good design is no more expensive than bad design, and that if good designs are not available for the man in the street the system which produces these designs must be undemocratic and therefore wrong." (175) The AFA responded by passing a resolution urging the recently established Federal Board for Vocational Education, which controlled the funds appropriated through the Smith-Hughes Law, to adopt the principle "that industrial art be given a prominent place in all vocational education supported by

this law." The AFA was convinced that "good design and the highest type of workmanship in American manufactures are absolutely necessary to enable the United States to hold a foremost place in the world's commerce." (92, 373) It charged the government that had recently passed the Smith-Hughes Act with the responsibility to set up industrial-arts schools comparable to the best ones in Europe. Furthermore, the AFA recommended that exhibitions and competitions be held to help manufacturers find talent and become familiar with the requirements of the trade in order to raise the standards of American industries to the level of those of the Europeans.

The Metropolitan Museum of Art, recognizing that American manufacturers had lost their sources of design, had already determined, at the urging of Richard Bach, to make its department of industrial arts directly useful to designers, producers, and dealers by encouraging them to use its collections as a source of inspiration. It instituted a series of annual exhibitions of American industrial art that ran from 1917 through 1931. The museum at first stipulated that a product submitted for exhibition must have been based on an item in its collections, but over the years this rule was gradually relaxed to allow products that had found their design sources elsewhere, and toward the end the exhibitions were broadened to include original designs. As a result, this series of exhibitions neatly brackets the shift in the design of manufactured products from imitation to innovation.

Walter Sargent, the director of the Department of Fine and Industrial Art at the University of Chicago, was convinced that although American designers had ingenuity they could not compete with the foreigners: "Because we had no comparatively stable aesthetic standards which we could modify and refine from year to year, we had to compensate by producing each season something entirely novel in order to meet an untrained but insistent popular demand. . . . It will not be sufficient to copy even skillfully, foreign designs. If we are to compete successfully we must cultivate originality." (196, 442) He laid out several principles in his paper to the 1918 AFA convention that suggested a proper base for design education in the United States. It should be realized, he stated, that new patterns and designs depend upon evolution and not servile imitation, that ideas and ideals may be drawn from museum study without abject copying, that nature can be used as a source for fresh design ideas, that manufacturing processes must be well understood and concepts must be capable of automatic repetition, and that designers could be good technicians without being actual workers in the trade.

Other educators, like James P. Haney, director of art education in the New York City schools, supported Sargent's proposals for training young Americans in the industrial-art trades. They endorsed his call for designers and his view of the philosophical goal that motivates designers: "Design . . . is not a luxury, but is based upon an inherent need; an elemental insistance that all constructed objects shall not only serve practical purposes, but possess also some beauty or distinction, a hint or symbol of something which is one step at least beyond utility." (196, 442)

By 1919, most of the American organizations concerned with design in one way or another, including the Art Alliance of America, the National Academy of Design, the Art-in-Trades Club, the Architectural League, and the National Society of Craftsmen, were working together under the title of the National Association for Decorative Arts and Industries toward common objectives. They acknowledged that the war had created a need for practical craftsmen and that Americans now had to be trained in the industrial arts. They recognized that the few highly trained foreigners still coming to the United States were invaluable in the building up of industrial arts. Moreover, they had learned that manufacturers had a need for product directors who would serve as the "architects" for a product, determine what general effect was wanted, and then call in designers to achieve that effect. It was felt that the designer should serve humbly, with self-repression more important than self-expression. The products would then be part of the new movement—the art that belongs to the present.

Despite such activities on the part of professional designers and the admonitions and pleas of educators for support for design training programs, it became evident as the war ended that Americans would be left behind. H. M. Kurtzworth, head of the Grand Rapids School of Art and Industry, published a study in mid-1919 reporting that in numbers of schools the European countries were far ahead. France had 32 design schools in operation, England had 37, Italy 24, and Germany 59, whereas the United States had only 18 by his count. He might also have pointed out that the foreign schools were supported by the governments as essential to the improvement of the trade and commerce of each country, whereas American design schools were private institutions that had no direct commitment to further the economic interests of domestic industry. In addition, Kurtzworth was concerned that the school programs that did exist were too involved in traditional academic theory and expression and did very little to prepare their students to earn a living. It must be pointed out that Kurtzworth's primary interest was in industrial-arts products such as furniture, textiles, tablewares, and the decorative and costume arts rather than in industrial-design products (hardware and machinery, domestic appliances and conveniences, communication instruments, business and industrial equipment, and vehicles for land, sea, and air). Neither he nor the other outspoken advocates of the industrial arts had noted that mass-produced utilitarian products had a unique identity that was deserving of attention on its own. Only one lonely voice in the *Magazine of Art* tried to point out that, since the machine was here to stay, it would be wise to study the peculiar requirements of design for machines.

The American manufacturers and merchandisers were not moved by all these attempts to establish a viable basis for American design in the industrial arts. No sooner had the armistice been signed than advertisements began to appear from merchandisers promoting the fact that their stocks of imported products had been replenished. Industry ignored the exhibitions that were being held to show what American designers could do.

In the summer of 1919, when an exhibition of
French industrial arts directed at buyers was held
at Hotel Pennsylvania in New York, the Art Alliance used the event to appeal that unless America acted promptly hope for American design
education would be lost and to announce a three-
year program for better "industrial design in
merchandise designed in America and made in
America by Americans" (177)—all to no avail.
Only one conclusion could be drawn: that the
industrial-arts industries in the United States,
despite the excellent work of the Arts and Crafts
community and the pressure from educators,
would (for a time, at least) prefer to use designs
that could be endlessly adapted from imported
examples.

Commercial Art
Discovers Design

Design is now studied in relation to its purpose. . . . Art as now taught is not an end but a means . . . we are at last beginning to teach art as an economic as well as an aesthetic factor.

Frank Purdy, 1921 (102, 208)

As the 1920s opened, there was still a polarization between those who believed that the Americans either could not or should not hope to develop their own design capabilities in the decorative and industrial arts and those who were convinced that the Americans must be awakened to the necessity of adding aesthetic value to their products if they were to meet worldwide competition. Some felt that, since the United States was peopled with immigrants, the resulting mixture of races was not conducive to the evolution of a national cultural identity. Many well-meaning Americans seemed to be caught in a web of European fashion and taste, fascinated with foreign expression and forever doomed to look abroad for cultural leadership. They were either caught in the avalanche of artifacts from the continent when "great decorators of note ravaged Europe for the spoils of the luxurious reigns of the Louis' and the splendid relics of the Renaissance" (178) or caught in the apologies of those who claimed, as Mrs. John Henry Hammond did, that she bought foreign things "not because I prefer foreign goods, or have necessarily greater confidence in foreign goods, but because in so many cases they have suited me better; their artistic merit is higher, showing the results of more highly trained designers and artisans." "If I could always find what I wanted in American goods," said Mrs. Hammond, "I would support American industry every time." (100, 38) Louis Tiffany contradicted her, saying that "the average American would rather bring back poor and thoroughly inartistic work from abroad than purchase domestic art in his own country." Tiffany expressed his frustration in attempting to convince manufacturers that they could benefit by supporting American designers: "Our manufacturers are entirely too commercial. We imitate rather than originate. . . . The artist must be as much a part of the business as the efficiency expert." (98, 196) Unfortunately, American manu-

facturers were slow in recognizing American creative talent and skill.

Tiffany and others believed that World War I had marked a turning point—that the American industrial arts, despite the general feeling that they were generations behind those of Europe, would begin to catch up. William Frank Purdy, president of the Art Alliance in 1920 and industrial-arts editor for *Arts and Decoration* magazine, was one of the strongest advocates of American industrial arts. He recognized that "whenever the art element was needed, it seemed the simpler and the surer course to import it ready-made—in the form of design, artists and craftsmen, or finished products—from Europe, although we had paid dearly for this privilege." (97, 27) By 1921 the *New York Times* was including the category of industrial design in its index, although the title was used primarily as a reference for the industrial arts. A strong sentiment had begun to develop for government-supported training of American designers.

One result of this trend was the launching in 1920 of a comprehensive study, under the direction of Charles R. Richards of the Cooper Union, of the state of the industrial arts and industrial-arts education in the United States and abroad. The study was administered by the National Society for Vocational Education with grants totaling $120,000 from the General Education Board of the federal government and the University of the State of New York. Although the study did not succeed in convincing the federal government to support industrial arts education, and the responsibility was thus left in the hands of private schools and a few state-supported academic institutions, it did reflect a shift in the traditional approach to education in the design arts.

The Richards study was directed primarily at those industries in which design exercises an important influence—the "art industries," in contrast to the "artless" industries. The art industries (some 510 of them, including textiles, costume jewelry, silverware, furniture, lighting fixtures, art metalwork, ceramics, glass, wallpaper, and printing) were asked whether they believed that they would benefit from an improvement in American design education. No industry rejected outright the concept of American-trained designers, but all expressed a reservation about the ability of young American designers to fill their special needs. The furniture industry feared that American designers would want to make original designs in violation of that industry's long-established custom of producing historical styles (preferably those of England). Silverware manufacturers were quick to point out that the high capitalization of their tools and dies discouraged innovative design, and that they preferred to employ highly trained technicians rather than designers. The general feeling of the industrial-arts industries was that, if any new ideas were to be introduced, they would have to be originated by architects and decorators in the specialized service of wealthy clients. They were suspicious of the intrusion of young American design ideas into their established markets.

The Richards study also noted that there were 274 schools of art listed in the 1920 *American Art Annual,* of which 58 were the most serious about the applied or industrial arts (although there was no clearly discernible interest on their part in finding places in industry for their graduates). The list included such privately endowed schools as the Maryland Institute, the Ohio Mechanics Institute, the Cleveland School of Art, the Otis Art Institute, Cooper Union, Pratt Institute, the School of Applied Art for Women in New York City, the Philadelphia Museum School of Design, the Carnegie Institute of Technology, the Rhode Island School of Design, the Rochester Athenaeum and Mechanics Institute, and the Skidmore School of Art. Other schools were associated with universities, such as the College of Industrial Arts at Denton, Texas, the Newcomb School of Art in New Orleans, the Art Department of Teachers' College of Columbia University, the School of Fine Arts of Washington University in St. Louis, and the College of Fine Arts of Syracuse University. Private art schools in the Richards survey included the California School of Arts and Crafts, the New Art School of Boston, the School of Fine and Applied Arts in New York, the Academy in Cincinnati, and the art schools of the Albright Art Gallery in Buffalo and the Art Institute of Chicago.

The list also included several public schools, such as the Massachusetts Normal Art School, the Milwaukee State Normal School, the Fawcett School of Industrial Art at Newark, and the Evening School of Industrial Arts in New York. In all, there were some 55 schools teaching industrial arts in the United States in the early 1920s. However, it seems that the graduates of these schools were no more interested in the industrial-arts industries than the industries were in employing the graduates. The schools were preoccupied with the lingering Arts and Crafts movement, and few, if any, looked to industry as a primary area of service. Nowhere was there any evidence of industrial design, as such, being considered of sufficient academic value to serve the "artless" industries—those manufacturing the essential products of everyday life.

The Richards study reported at length on the need to attract promising young people to design rather than to the fine arts:

One consideration that affects the quality of American youth entering upon applied-art education is the essentially modern quarrel between the fine arts and the applied arts. The idea that the fine arts as represented by painting and sculpture are something superior to the applied arts and that their practice is a matter of greater dignity is an attitude that persists tenaciously. There is still a vast difference in the appeal to young persons as between the career of a painter or sculptor and that of a designer. Even in schools where courses in both the fine and applied arts are given, the school authorities are very often found influencing talented students toward painting and sculpture and away from the study of industrial art. We are only slowly coming to recognize the true meaning of the applied arts in our national life. We are only gradually coming to recognize that art is fine not because of a particular medium, but when the expression of line, mass and color is fine and beautiful, whether this be in a painting or a rug, and that art is not fine when this expression is poor and commonplace, whether the medium be sculptured bronze or a piece of furniture.

To obtain better student material in our art schools we also need not only higher material rewards for designers but a more recognized and dignified status. With us the designer has practically no status other than that of a worker in the industries. In Europe he is regarded as an artist and occupies a dignified position in the community. (74, 493)

The concern expressed about appropriate recognition of the designer in this last point in the Richards study touches once again one of the most sensitive issues affecting the quality and potential of design. To some degree this lack of recognition was blamed on the fact that the work of the designer was considered to be an anonymous part of the total process of production. Yet it seems obvious that if spirited designers were to be attracted their work should be publicly recognized, whereas anonymity would serve the notion that designers were colorless slaves of the industrial system.

Another factor that kept designers anonymous was the predilection of many industries for the fashion of the moment. In 1921 Matlack Price, in the magazine *Arts and Decoration,* put his editorial finger on anonymity as one of the reasons why the industrial arts were not attracting outstanding talent: "The anonymity of nearly all our designers in the field of industrial art has had many disadvantageous results. The public has been denied the opportunity and pleasure of recognizing and enjoying the work of any individual designer, and many men of real ability have hesitated to become industrial designers because of the consequent loss of their identity." (107, 166)

It was also proposed that the better artists, architects, and painters preferred to remain anonymous when they provided designs or models for products to be manufactured because they feared that their reputations as fine artists might be put into jeopardy if they were to be identified with commercial products. In other countries the designers of manufactured products were proudly listed by manufacturers, but American manufacturers preferred to keep their designers under wraps. Perhaps they were concerned that if designers gained public recognition they might become more independent and demand more money and other favors.

Anonymity still prevails in American design, although a few products, mostly in the area of high style and high-fashion furnishings, have brought their designers rewards and public recognition.

The rationale supporting anonymity maintains in part that in industry no design can be the work of one person and that, therefore, the list of persons who would have to be given credit would be much too long to be practical. Yet if a design results in a successful product it attracts many "parents," while a bad design remains an orphan, its parentage conveniently concealed. It is conceivable that there is so much bad design around because accountability for it is not mandated by either professional ethic or public law. Until the time when the perpetrator of an aesthetic horror or fraud is called to account, one may expect that anonymity will protect the incompetent designer to the same degree that it now often robs the able designer of just compensation for an eloquent product.

Art museums in America had maintained for years that they had a unique role to play in preserving the fine and industrial arts. They had been conceived and endowed as showcases for the cultural acquisitions of the affluent, "to inspire not only the newly-hatched millionaire, but also the moneyed day-laborer." (180) Now they defined a new role for themselves as a design resource for industries by offering access to their collections to "all industries in which artistic design is in any way a dominant element." (101, 116) Their new goal had already been pointed out by the Metropolitan Museum of Art's exhibitions that were initiated in 1917. Richard F. Bach of the Metropolitan expressed the museum's position: "There is fertile virgin soil for the art museum offering direct as well as subtle lines of art influence by which, properly used, museums may bind themselves forever to the most intimate feelings of the people, reaching them through their home furnishings, their utensils, their objects of personal adornment, their clothing." (101, 116) Bach was proud of the title "The Workbench of American Taste" that had been bestowed on his department in the museum, and saw nothing unnatural in the concept of industries as "great capitalizers" of art. He saw the museum's role in the use of art as an element in business not as a sign of art's degradation, but as a sign of art's progress. Bach was convinced that the Metropolitan shows were "a direct reflection of trade conditions, recording as truly as the money market itself the ebb and flow of prosperity, the ascendancy of the new rich, ill-begotten fads created out of hand by scheming producers, unemployment, strikes and the devious ways of modern selling," and the Museum as "a quietly effective teacher" encouraging in stylemakers "the tendency to go back to the strong and unpolished periods of art for inspiration." (183)

Museums seemed to be certain in the 1920s that if they offered their collections to industry as sources of inspiration, manufactured objects would somehow be imbued with the vitality and quality of the original objects they displayed. The Metropolitan boasted that its exhibits had rendered good service to those in search of design ideas. Florentine glass, Italian gesso picture frames, and medieval armor were pointed out as having "provided" ideas for fabrics. A wallpaper manufacturer "found" his ideas in ecclesiastical vestments, an Athenian jar "inspired" the form of a modern cosmetic jar, and a paper soap wrapper "saw" its beginning in antique snuff boxes. Bach was ecstatic about the success of the Metropolitan shows: "The real truth of our progress lies in the designs which are the result of what may be termed the inspirational use of the collections—when a lamp manufacturer gets ideas from Cellini bronzes or Greek mirrors—this means progress. When a neckwear manufacturer studies armor; or a tile designer studies miniatures, we may safely say the clear light of a new day is dawning in American design." (103, 302)

Some museum directors believed that they should play an even stronger role in the design of products than that of a source of inspiration. F. A. Whiting, director of the Cleveland Museum of Art, writing in the December 1920 issue of the magazine *Art and Decoration* on the subject of "The Museum and the Industrial Designer," expressed discouragement with the misuse of museum collections by designers of poor taste and recommended that the museum should reserve the right to approve each design before it was reproduced.

Only one museum director, John Cotton Dana of the Newark Museum, dared to raise his voice against the use of museum collections as a source of ideas by manufacturers. He was convinced that rather than encourage American designers they tended to make them believe that they could not be artistic or creative without the museums' help. "Could the money," he wrote, "spent in one year by art museums and wealthy collectors in this country in patronizing the modern and ancient art of other countries be spent in a carefully directed patronage of art movements in this country, we would see here the opening of a very good example of art 'naissance.'" (101, 117) In order to stimulate designers and manufacturers, Dana repeated the prewar exhibition of industrial arts from the Deutscher Werkbund, with the aim of "ennobling and beautifying everyday life by creating artistic surroundings . . . down to the simplest kitchen utensil." (139, 332) In 1920, the Art Institute of Chicago supported Dana's point of view by staging an exhibition of the work of the Wiener Werkstatte, with the strong support of the Art and Industries Association of Chicago. However, these exhibitions, too, may be criticized in retrospect as attempts to bring the light of originality to the Americans from abroad.

In 1921 the author of an unsigned *New York Times* article on the Metropolitan Museum's Exhibition of American Industrial Art expressed the cautious hope that "great would be the joy if something American and modern, something all our own, and for that reason legitimately dear to us, should prove of finer grain, of stronger fiber, of more engrossing beauty, than the Florentine or Egyptian or Versailles mode." (183) In the same year Charles Richards scored the museums for their narrow vision by noting in his report on the industrial arts in America that museum collections were "acquired originally for pleasure and profit" and expressing the hope that "art for people's sake [should be] the motto of the American museum." (74, 436) But manufacturers did not use the museums as a source for quality, and museums did not recognize the cultural value of contemporary products. Richards felt that the museums should be awakened to the necessity of deciding the part they should play in modern industrial life. The museums had pledged to enlighten the public and to stimulate better designs by serving as inspirational centers, but they may actually have diverted and even retarded the evolution of indigenous American design.

The dearth of native industrial-art designs in the United States and the pent-up desire of American industrial-arts manufacturers for foreign ideas were not lost on the Europeans. It was in the early 1920s that most of the foreign designers, architects, artists, and artisans who were to form half of the first generation of industrial designers in the United States emigrated there. One exception was the talented young German Kem Weber, who had come over in 1914 to install his professor Bruno Paul's exhibit for Berlin at the Pan American exposition and had been stranded in the United States by the outbreak of the hostilities in Europe. Weber (1889–1963) eventually became an American citizen, and for many years he was a dynamic force in furniture and interior design on the West Coast. Most of the other European designers who came to the United States, though they were disheartened by the postwar depression and political crises in their homelands, were not refugees. They came to take advantage of new opportunities at a time when America was welcoming adventurous and high-spirited foreigners. Like Weber, the majority of them had been trained primarily as architects or artisans. They brought with them the moderne style, which they applied first to decorative-art furnishings, fabrics, wall decorations, and furniture and later to other special-assignment or manufactured products. In America the émigré designers found a ready outlet for their work, primarily through exclusive shops promoting foreign products and talents to the more daring members of the affluent class who were enamored with the modern art movement in Europe.

Among the designers who came to the United States in this first postwar wave were George Sakier and William Lescaze from France, Paul Frankl and Joseph Urban from Austria, Paul Laszlo from Hungary, and Gustav Jensen from Denmark. They were followed in the mid-1920s by Peter Muller-Munk and Rudolph Koepf from Germany. Three notable early émigrés were Joseph Sinel from New Zealand, Raymond Loewy from France, and John Vassos from Greece.

Kem Weber. Courtesy of George Jergensen.

Joseph Sinel. California College of Arts and Crafts, Oakland.

These three men, who helped to pave the way for industrial design in the United States, were neither architects nor artisans. Rather, they were primarily artist-illustrators whose association with advertising illustration led them into the design of packaging as well as of the products themselves.

Joseph Sinel (1889–1975), who emigrated to America in 1918 from New Zealand by way of Australia and England, once made the bold claim that he "was the first established industrial designer in the United States." (43, 58) He may well have been in the strict sense of the term, because by 1921 he was taking on product-improvement assignments from the advertising agencies that he worked for and had already taken out design patents for products. However, Sinel was careful to point out that he did not invent the term industrial design, and was embarrassed by persistent inferences that he had. Although most of Sinel's early training had been in lithography and illustration, his assignments in America broadened quickly to include virtually every facet of what is now commonly called industrial design. His 1923 book on trademark design, *A Book of American Trademarks and Devices,* anticipated the specialization that is known today as corporate-identity design by advocating that a good trademark should help to establish an appropriate image of a corporation in the public mind. Sinel's rules for trademark design pointed out that a good mark must be simple yet unique and flexible enough to suit all anticipated applications.

Raymond Loewy (b. 1893), a French engineer whose name was to become strongly identified with industrial design in the United States, left France in 1919 with the equivalent of $40 in his pocket to come to the United States to work with his brother Georges, who had preceded him. En route, Loewy, who had been drawing all his life (as a child he had filled notebooks with drawings of automobiles and trains), produced some fashion sketches that prompted his fellow passenger, the British consul general, to give him a letter of introduction to Condé Nast, then editor of *Vogue* magazine. After his first job as a window dresser for Macy's, Loewy was asked by Condé Nast to do fashion sketches for *Vogue.*

Trade card designed and used by Joseph Sinel in New Zealand around 1912, prior to his emigrating to Australia, to England, and then to the United States in 1918. California College of Arts and Crafts, Oakland.

Raymond Loewy. Fortune, February 1934.

Advertisement in high fashion of the day, designed by Raymond Loewy. Vanity Fair, February 1927.

From this beginning, Loewy went on to advertisements in the fashionable French Art Moderne style for various companies and department stores, costumes for Florenz Ziegfeld, and uniforms for elevator operators at the Saks 34th Street store before he got his first assignment in product design in the late 1920s.

The adventurous John Vassos (b. 1898) tested his talents early in life on his father's newspaper in Constantinople by drawing political cartoons of Turkish officials. Forced to leave Turkey in 1915, he joined the British Naval Support Systems, for which he served in World War I. In 1919 Vassos came to the United States to study art and illustration with John Singer Sargent in Boston and later with Sloan and Bridgeman in New York. His distinctive style of illustration began to earn him important commerical-art commissions for magazines such as *Harper's* and *The New Yorker.* His first industrial-design assignment, in 1924, was to design a new face-lotion bottle for a cosmetics company. Vasso's bottle became popular as a liquor flask during Prohibition—people "emptied the face lotion, washed it well, and filled it up with gin!" "By the way," noted Vassos, "I also introduced the plastic screwtop. The sales went up 700%." (219) Vassos managed to find time to paint murals and to write and illustrate books. Later he figured in the establishment of the first professional design organization in the United States (the Industrial Designers Institute) and, with the Persian émigré painter Alexander Kostellow (1898–1956), developed one of the earliest complete programs in industrial-design education.

The productive capacity that had been developed in the United States to meet wartime needs was now being redirected to meet the daily needs and desires of a population that had just passed the 100 million mark. Moreover, since the number of kilowatt-hours of electricity available to each citizen had doubled since 1915, an enormous market was opening for electrical appliances—not only the smaller ones that had been available before the war, but now, in the 1920s, for the electric ranges, refrigerators, and washing machines that were just coming onto the market. Kitchen cabinets were also becoming appliances.

John Vassos. Fortune, *February 1934.*

John Vassos made his reputation as an illustrator and a muralist with a socio-political message before he turned to industrial design. One of his most enduring works has been the book Phobia, *from which is taken this illustration of "mechanophobia," the fear of machinery.*

John Vassos's recyclable cosmetic bottle for the Armand company became popular as a hip flask during Prohibition. John Vassos.

*The Maytag company
carefully refined its electric washing machine into
a typeform before the rush
to industrial design. Maytag Company.*

*The kitchen cabinet sys-
tem manufactured by the
Sellers company in 1923
was, along with others,
a step toward the com-
pletely built-in kitchens
that were to take over in
the next decade.* Roaring
Twenties, *November 1923.*

The first major electrified appliances were essentially conceived as invented forms whose primary purpose was to function efficiently and safely. Such appearance details as were considered were relegated to legs, support brackets, and hardware. However, the General Electric Company had demonstrated its concern about the appearance of its products in 1912 by assigning J. W. Gosling of its Illuminating Laboratory in Schenectady to design the decorative lighting equipment for the 1915 San Francisco Fair, and in the early 1920s the company showed its attention to a developing market for consumer and industrial products by establishing a committee on "product styling," of which Gosling was a member. This group was concerned with the appearance of products ranging from fans to motors, generators, turbines, and locomotives. Soon GE, like other electrical companies, moved into domestic appliances and entertainment and communication products.

The primary problem that faced manufacturers was not production, but rather how to put information about their products into the minds of prospective customers. In a happy coincidence, the advent and commercialization of radio provided the answer. A product mentioned on the air during an evening radio program would be demanded by thousands in stores across the country the very next morning. And handsome illustrations of products in mass-produced and now widely distributed magazines helped to whet the public's appetite all the more. It quickly became apparent that the appearance of the product in an advertisement would be an important element in its public acceptability, and this placed the advertising agency and its artists in the position of having to make certain that the product being promoted was as attractive as it was useful. The public, flattered by all the attention being lavished upon it by manufacturers and the media, seemed unaware that it was paying for the privilege of being sold and assumed that it was entitled to be entertained by the media as well as to receive good-looking products as cultural prizes for its marketing dollars.

At the 1921 convention of the American Federation of Arts in Washington, Leon W. Winslow, a specialist in art education at the University of the State of New York, observed that art was becoming more important to both manufacturers and consumers. Whether or not the product had inherent artistic qualities was, he felt, essential to the proper advertising of any product as well as a controlling factor in many industries where design was involved in the construction as well as decoration of the product. Industry, Winslow concluded, was interested in art primarily from a commercial point of view and was thus on the hunt for artists who could produce salable products. In the same year, Alvin E. Dodd, secretary of the National Chamber of Commerce, expressed the same theme from the viewpoint of business in a talk to the annual conference of the Eastern Arts Association. He was convinced that art was one of the chief influences in competition: "The auto with the best lines, the advertisement with the best lay-out, windows with the best display, draw the most business. . . . The use of art to increase sales is a direct appeal to the appreciation of art by the businessman." (105, 27) However, Frank Alvah Parsons's contemporaneous essay "The Relation of Beauty to Fashion" warned that, though the desire for beauty was a universal instinct, in practice it could be "tyrannical" in its demands. He recognized the ephemeral quality of beauty as aiming to satisfy human desires for change, distinction, or status. Its appeal to vanity encouraged invention and commercial exploitation, and thus fashion stimulated competition and affected supply and demand.

Coincidental with the realization that design had an important role to play in the competition for the consumer's attention was the emergence of business as an important academic discipline. The first course in business had been pioneered as early as 1881 by the University of Pennsylvania, but it wasn't until the 1920s that business programs were introduced into the curricula of many major American academic institutions. By 1925, Stanford University was awarding a doctorate in business administration. The promotional arm of business became known as advertising and was recognized as "the ignition system of the

A 1921 advertisement for an electric range by the Hughes-Hotpoint division of the Edison Electric Appliance Company, which eventually became part of General Electric. This range is strictly a mechanical product made palatable to the consumer with a scroll bracket on the shelf and cabriole legs. Good Housekeeping, March 1921.

The General Electric trademark. The company has resisted pressure to modernize this mark. Saturday Evening Post, September 1923.

Hotpoint HUGHES

E L E C T R I C   R A N G E

The initials of a friend

You will find these letters on many tools by which electricity works. They are on great generators used by electric light and power companies; and on lamps that light millions of homes.

They are on big motors that pull railway trains; and on tiny motors that make hard housework easy.

By such tools electricity dispels the dark and lifts heavy burdens from human shoulders. Hence the letters G.E are more than a trademark. They are an emblem of service—the initials of a friend.

GENERAL ELECTRIC

The General Electric
"monitor-top" refrigerator
was manufactured with
only minor changes in
styling from 1927 until
1935, when the compres-
sor was moved to the
bottom. Cabriole legs,
arts-and-crafts style
hinges and latch, and
a unique form made the
original a classic type-
form of the era. The later
squared-off hardware
reflected the modernist
style. General Electric
Company.

Although Herbert Hoover,
Secretary of Commerce
in 1922, could not believe
that the American public
would put up with "adver-
tising chatter" on the
wireless, the public was
enraptured with radio and
heeded its appeals to buy
in return for the "free"
programs that it offered.
C. D. Gibson drawing;
Life, 1922.

WHAT ARE THE WILD WIRES SAYING?

economy, the dynamo of mass dissatisfaction and the creator of illusions in a most materialistic world." (67, 15)

It is logical that advertising should have been one of the two elements that established industrial design in the United States. It was emerging as the most sensitive of all of the business forces, offering financial rewards and larger opportunities to artists to become a part of the technological economy. Modern printing methods, high-speed presses, and multicolor advertising gave magazines a great advantage over traditional poster promotion. Mass-produced magazines brought handsome promotional material into the home, where the promise of the product and its beauty could be pondered at length in privacy. "The forces," wrote Ernest Elmo Calkins, "that are making our fast-paced, bright-colored, sharply defined civilization are producing our modern art. It is appropriate that modern art should enter business by the door of advertising." (121, 153)

Ernest Elmo Calkins must be credited with having had the vision to see that success in mass production depended upon mass promotion and that, therefore, packages as well as products would have to be redesigned if they were to attract and hold the attention of a vast buying public. Joseph Sinel claimed that the Calkins and Holden firm was the first advertising agency in the United States to offer this creative service, and it was probably from Calkins that Walter Dorwin Teague caught the spark that he was to blow into the flame of industrial design later in the 1920s.

Walter Dorwin Teague (1883–1960) was involved in design as early as 1903, when he was studying at the Art Students' League in New York and supporting himself partly by lettering signs. He found freelance employment drawing shoes and neckties for mail-order catalogs before taking full-time employment with the Ben Hampton Advertising Agency. In 1908, when Teague's employer Walter Whitehead went over to the Calkins-Holden agency, he took Teague with him. Teague, like Sinel, credits Calkins with having taught him what he knew about making a successful business out of art.

Walter Dorwin Teague.
Fortune, *February 1934.*

Teague border and illustration from a 1927 Literary Digest *publication.*

This dramatic black-and-white illustration for an automobile advertisement by Teague is typical of the strong high-style technique that was suitable for the publishing technology of the day. Art in Industry, *1923.*

Two out of every Three
DIGEST *Subscribers have Telephones*

By 1911, Teague was operating his own art agency and specializing in the decorative typographical borders of the day. Teague's dignified, rather classical borders set a style in the industry that came to be identified with his name. From borders he expanded his services to the design of full advertising pieces, including illustrations in which, to please his clients and attract the attention of sympathetic purchasers, he must certainly have emphasized and enhanced the finer qualities of the products. The 1921 *Art Annual* lists Teague as a designer of advertising for such companies as Community Plate, Phoenix Hose, Arrow Collars, and Adler-Rochester Clothes. By 1922 his advertising designs were being included in the annual industrial-art exhibition at the Metropolitan Museum of Art. When the editor of the magazine *Art and Decoration* called particular attention to the fact that such promotional styling "makes beautiful a thing of primarily utilitarian character," he was referring to the high quality of Walter Dorwin Teague's layouts and advertising. (106, 174) As the demand for decorative borders and the like began to decline in the mid-1920s, Teague found himself drawn to the new spirit of design that was sweeping France. He returned from a visit to Paris determined to do no more borders but, rather, to concentrate on product design.

Egmont Arens (from his 25 Years of Progress in Package Design).

Egmont Arens (1888–1966) was sent by his family to New Mexico to recover from tuberculosis, and began his career there in 1916 as sports editor of the *Albuquerque Tribune-Citizen.* After a year he moved to New York, where he purchased and operated the Washington Square Book Store and established (in 1918) the Flying Stag Press as a printer of magazines. Arens's interest in the arts led to the editorship of *Creative Arts* magazine and then of *Playboy* (the first American magazine devoted to modern art). Later he became editor of *Vanity Fair,* one of the leading style and fashion magazines in the country. From these beginnings, Arens went on to become a leading packaging and product designer. Also, from 1929 to 1933, he headed the industrial styling division of the Calkins-Holden Advertising. He coined "consumer engineering" as the title of his philosophy of matching products to public interest and demand. Once again, one can sense the influence that Ernest Elmo Calkins had on the practice of industrial design in the United States.

Donald Deskey (b. 1894) did not consider himself a pioneer industrial designer, yet he was one of the first Americans to find a proper and profitable connection between modern art, advertising design, and manufactured products. After an adventurous youth on the West Coast and sporadic art training, he took a job in 1920 with an advertising agency in Chicago at $15 a week. He went to New York in 1921 to work for an advertising agency there, only to be fired because his work was considered to be too modern. His response was to open his own art agency, and soon it was paying him the princely salary of $12,000 a year. The restless young Deskey was convinced that the most exciting art was being done in Paris, so he closed his office and went there. He returned to the United States in a year to head the art department at Juniata College in Pennsylvania and to refine his theories on art and design, but within a short time the runaway success of the Decorative Arts Exposition in Paris drew him back for a second and perhaps more telling exposure. In a way, Deskey's decision to go abroad in order to satisfy his curiosity and perhaps expand his artistic sensitivity was not so very different from the thinking of those manufacturers who chose to be inspired by foreign ideas rather than nurture native ones, or from the position of those museums that offered foreign treasures as sources of inspiration to benighted American designers. Upon his return Deskey combined his artistic ability with product design by producing decorative screens and other furnishings that found favor with architects working in the Art Moderne style. In addition, his products found their way into stage settings as the epitome of modern art in the theatrical world.

Lurelle Guild (b. 1898) also showed an early interest in the theater (as an actor). He received a degree in painting from Syracuse University, where he showed signs of his unique ability to combine aesthetic sensitivity and business acumen. As an undergraduate Guild maintained an off-campus studio where, with other students

Donald Deskey. *Fortune,*
August 1943.

Deskey's bold use of mod-
ern materials such as cork
and linoleum for interiors,
and his use of traditional
materials such as rattan
to create modern prod-
ucts, helped him acquire
major clients. Annual
of American Design
(AUDAC), *1931.*

that he hired, he produced advertising art and design ideas for local agencies and industries. As he wrote to the author,

A year after graduating from Syracuse in 1920, I was selling covers to *House and Garden* and illustrated articles to the *Ladies' Home Journal, Delineator, Pictorial Review,* and other periodicals. The editorial acceptance of my drawings led to illustrating advertisements and by 1923 I found that the advertisers also wanted me to design their products. In 1926 I developed a textured embossed linoleum for the Armstrong Cork Company. Although I had had many appearance design patents, this was my first mechanical design patent. At that time I used the title of 'The Guild for Industrial Guidance' for my business and in 1928 changed it to 'Guild of Industrial Design, and still later to 'Lurelle Guild Associates.' (213)

Guild (who to this day holds the greatest number of design patents for products) also pioneered the techniques of testing market acceptance by interviewing consumers—he would often tour neighborhoods with a truckful of products.

Lurelle Guild. Fortune, February 1934.

The theater shares with advertising the distinction of having been the early source of industrial design in America. In the make-believe atmosphere of the stage, just as in advertising, success often depends upon the ability of the designer to translate the "client's" (the playwright's) intentions into a believable experience for the public. Theater thrives upon sensitivity, imagination, and daring; so does advertising. Furthermore, experience in the theater is, like advertising and the products that it normally promotes, transitory. Designers of theatrical settings were able to make a convenient transition, first to the design of show windows and product displays and then to the challenge of bringing the actual and apparent quality of the products displayed up to the quality of the exhibit and the enhanced anticipation of the public.

Norman Bel Geddes (1893–1958), of all the emerging American industrial designers in the early 1920s, stands out as the first to have felt the cultural surge of the twentieth century. This dynamic and voluble showman, the P. T. Barnum of design, who successfully combined careers in the advertising and theatrical arts, was the catalytic designer of his generation. For over thirty years a surprising number of important American designers acquired their first exposure to design practice in Bel Geddes's studio and workshops. He came closer than any of his contemporaries to the ideal of the Renaissance man, as a painter, an illustrator, a graphic designer, a conceiver of window displays, exhibits, theatrical settings, and extravaganzas, an inventor, an architect, and ultimately a product planner and an industrial designer.

Before 1920, Bel Geddes, a high-school dropout, had tried unsuccessfully to acquire formal art training at the Cleveland School of Art and the Chicago Art Institute, worked for Aline Barnsdall's Little Theatre in Los Angeles as a designer-technician, and served as illustrator and art director for the Peninsular Engraving Company in Chicago and then for the Barnes-Crosby Advertising Agency in Detroit. In 1918 he landed in New York as a theatrical designer. His "star" status was quickly ensured by two set designs. The first was his innovative and dramatic 1923 concept for staging Dante's *Inferno,* which, although realized only in model form, attracted enough attention in the threatrical world to land him the assignment for the second, staging *The Miracle* for Max Reinhardt. For this, Bel Geddes transformed the entire interior of the old Century Theater in New York into a Gothic environment and seated the audience on crude benches before the cathedral facade as participants in the performance. Thus, early in his professional career Bel Geddes was able to demonstrate the drawing power of grand spectacles and soaring design concepts as well as his lifelong conviction that the worlds of fact and fancy are often inseparable.

In retrospect, one American industrial designer seems to have been endowed with the deeper and more humane values (concern for the consumer and for the fitting of products to people) that have been claimed by the field of industrial design. Henry Dreyfuss (1904–1972) quit high school in a huff because his 100-percent grade on an art exam was lowered to 98 on the grounds that perfection in the subject was unattainable. He finished his education at the Ethical Culture School, where under the progressive educator Felix Adler he was indelibly impressed with the

Norman Bel Geddes.
Fortune, *February 1934.*

Bel Geddes designed
posters and advertise-
ments between 1910 and
1920, including this one
for Bagley's Buckingham
Tobacco in the bold style
of the era. Hoblitzelle
Theatre Arts Library, University of Texas, Austin.

Bel Geddes designed the
settings for a staging of
Dante's Divine Comedy.
This photograph shows
a model of the spectacle,
which was to be pre-
sented in a theater de-
signed by Bel Geddes in
the ancient Greek style.
Hoblitzelle Theatre Arts
Library, University of
Texas, Austin.

philosophy of a liberal education based on natural capacity, individual difference, and cultural necessity. Dreyfuss dedicated his career to proving that people were more important than products. This wise and scholarly man managed to practice industrial design with a sense of responsible service to both client and consumer.

Dreyfuss was—and, for many, remains—the conscience of the industrial design profession. Yet when he joined the staff of Norman Bel Geddes in 1923 as an assistant to work on sets for *The Miracle,* he seemed to be attracted more to the make-believe than the real world. From this he went on to design stage pieces and backgrounds for the Strand and other theaters and decorative schemes for the Roseland dance hall. In 1927, when he was in Paris, Dreyfuss received a letter from Oswald Knauth of the R. H. Macy Department Store in New York asking him if he would consider a new kind of job: "Would I pick out Macy merchandise that lacked appeal and make drawings in the form, shape and color I thought would sell better? The drawings would be submitted to the manufacturers, who would be expected to revise their products accordingly." Wrote Dreyfuss: "I took the next boat home." (27, 15) Back in the United States, Dreyfuss made a study of the store's merchandise and concluded that the best way to improve manufactured products was to work directly with the manufacturers rather than to second-guess them afterwards. Macy's considered his proposal wise, but was unable to take it up. From this experience, however, Dreyfuss determined to become an industrial designer. He opened his first office in 1929, and for a while he combined small industrial-design assignments with the design of stage settings for big Broadway musicals. In a way Henry Dreyfuss managed to find a balance in employing the imagination and fancy of the stage and the logic and dedication of design. "Joe" and "Josephine," cartoon surrogates for the public, appeared on his drawing board and led him through a long career of service.

Russel Wright (1904–1976) also began in the theater and then turned his sharpest attention to improving the quality of products for the domestic environment. As a young man he spent the

Henry Dreyfuss. Fortune, February 1934.

summer after his sophomore year at Princeton University with a friend in the art colony at Woodstock, New York, designing a series of papier-mâché circus animal costumes for the summer festival. Instead of going back to school, Wright took a portfolio of his work to New York City to show to theater designers including Robert E. Jones, Lee Simonson, and Norman Bel Geddes. Bel Geddes hired him as an assistant on sets for *The Miracle* and introduced him, as he had Dreyfuss and others, to the challenge of capturing the public's imagination through design.

Wright's summer experience at Woodstock (where he met his future wife and aesthetic guide, Mary, while she was studying sculpture with Archipenko) gave him the idea of making and showing small plaster models of his circus animals to Rena Rosenthal for sale in the New York gift shop that she and her husband Rudolf had set up (at the urging of her brother Ely Jacques Kahn) to market Art Moderne products. The shop, which may have been the first "contemporary" shop in the United States, became an outlet for the products of European designers and craftsmen. Mrs. Rosenthal was intrigued by Wright's sculptured pieces and convinced the young American that she could market them if they were cast in metal. Wright went on to make clever caricature sculptures in uniquely appropriate materials, including Herbert Hoover in marshmallows and Greta Garbo in blown glass, that were widely publicized in such magazines as the *New York Review.* From these experiences Russel Wright, the consummate craftsman, found his way into a career devoted, as he once said, to "humanizing functional design." (220) Like his colleagues, Wright believed that the practice of design should cut across traditional artisanal preoccupation with a single material, process, or product. Unlike his contemporaries, however, he found his métier in the world of decorative and useful products for the American home, and he devoted his career to an unremitting search for domestic forms that were uniquely American in character.

There were other American designers who entered the practice of industrial design from the promotional side. John Alcott, of Boston, began as a display manager for a department store in the early 1920s and then went on to design floor coverings before interesting the Massachusetts Department of Commerce in providing active support for those designers who wished to serve industry. Robert Sidney Dickens painted advertisements on the stage curtains of a burlesque house in Gary, Indiana, before moving up to the position of assistant window trimmer of an Army and Navy Store and the W. T. Grant Company and eventually establishing one of the most successful package-design offices in the United States. Others were drawn into the new profession from a broader range. Thomas Lamb began as a textile designer. Harley Earl and Alexis de Saknoffsky were custom automobile body builders. Harold Van Doren was a museum director. Jack Morgan was a haberdashery salesman. None of the American designers were architects or engineers, nor were any of them schooled in the European academies that were later to be given credit by some for the origin of industrial design in America. Rather, in one way or another, they were all drawn into the vacuum for industrial design that existed in the United States in the 1920s.

The Rejection of
Art Moderne

*. . . every effort should be made to encourage
the industrial designers now at work and to
enable the rising generation of artists to carry
on their studies until our achievements in the
practical forms of art not only equal but surpass
the achievements of other nations.*

William L. Harris, 1922 (185)

In 1920 the magazine *Arts and Decoration,* which
had long been the leading American supporter
of arts and crafts and the industrial arts, acknowledged that it should no longer deal with
the finer arts only, "but also with those accessories to life by which we express our individuality and indulge our personal tastes." (99, 28)
Editor Matlack Price pointed out that the broadening of design was attracting artists and artisans
from many fields and that designers were inter-
preting the term *design* as encompassing a
broad area of activity concerned with creating
objects that were beautiful, expressive of their
purpose, and practical in construction and manufacture. For a short time it appeared that American design would take off on a course of its own,
away from dependence upon European fashion,
toward a recognition of functional needs and
mass-production technology. Wrote W. F. Morgan
in 1920: "It is gratifying to note that a change
of sentiment is taking place. The average citizen
is beginning to understand that an article may
be useful and beautiful at the same time. . . . we
are realizing that there is really no essential
distinction in artistic character between the com-
monest household objects and the rarest productions of artistic genius." (100, 38)

The promise of expanded opportunities for artists
and designers brought new life into older organizations such as the Art-in-Trades Club (founded
in 1906), the School Art League (1911), and the
National Alliance of Art and Industry (1912) and
stimulated the establishment of new organizations such as the Gift and Art Association (1921),
the Association of Decorative Arts and Industries
(1920), and the American Institute of Industrial Art
(1922). Though most of these organizations have
long since disappeared, they served to point out
that conscientious designers, manufacturers,
merchandisers, and educators had recognized

the importance of design and art in manufactured
products, whether decorative or utilitarian or both,
and were searching for means of developing a
fruitful relationship with one another and advancing public taste and interest in better design.

The National Alliance of Art and Industry became
particularly active in 1920 when, with William E.
Purdy as its president, it sponsored meetings and
competitions to promote the industrial arts in
America. The Alliance was especially proud that
objects that had won some of its competitions not
only had been sold in Paris but had also been
passed off as coming from Paris to Americans
who would not have bought the products otherwise. Also in 1920, the Alliance formed an Industrial Arts Council, comprising officials of business,
design, and industrial organizations, to explore
the common bond that they had all found in design. As one of its first actions the Council proposed that an Art Centre be established as an
umbrella "organization to which artists might look
for aid in placing designs for manufactures, to
which manufacturers might look for appropriate
designs, and to which both manufacturers and
artists might go for the kind of instruction necessary to make industry hospitable to art and art
appropriate to industry," recognizing the fact that
it was not easy "to build an efficient bridge over
the chasm that has widened between modern art
and modern industry in this country." (182)

More than 100 separate organizations, ranging
from the Architectural League through the Art
Alliance of America, the Art Directors Club, the
American Institute of Graphic Arts, and the
Society of Illustrators to trade organizations and
craft societies of every type, joined the Art Centre
in the hope that the encouragement of the industrial arts among young Americans would make
future importation of European art goods unnec-
essary. For all of these artists and designers the
Centre, under the direction of William L. Harris,
held conferences and lectures, staged exhibits,
and maintained a bureau for placing designers
in the various trades and industries in its two
houses in Manhattan. Mrs. Ripley Hitchcock,
chairwoman of the executive committee of the Art

The symbol of the Art-in-Trades Club was evidence of the careful position that it attempted to take between traditional and modern influences in design. Club yearbook, 1922.

Alliance, in emphasizing her conviction that all of the arts in industry would have a place in the Art Centre, offered the following statement: "Americans must realize that an artistic garbage pail will appeal more strongly to the housekeeper than an inartistic one, even at a slightly increased price, although Americans will soon find that the artistic things are cheaper in the long run." (181)

The Gift and Art Association was founded in response to the remarkable growth of exclusive shops, both independent ones and those that were parts of department stores. The success of these shops seems to have been due largely to the higher wages that for the first time provided people with disposable income with which they could purchase things to brighten their homes, and also to the fact that magazines and newspapers were featuring home decoration as never before. It was also proposed that soldiers had come home from France with positive ideas about more artistic furnishings in the home. Consequently, industry recognized the value of more artistic products, and artists who had scorned commercialization now seemed more willing to design objects to be made by trained craftsmen. Though some artists were fearful that any attempt to make art a paying proposition would lead to standardization, the principal evidence seemed to be that art would play a greater part in the selling of objects and there would be fewer reservations about designing useful as well as artistic products.

The emergence of design as a distinctive calling in the United States seems to have been directly related to the growth of mass production as a respectable source, not only for the necessities, but also for the niceties of democratic life. Although some accused the machine of being the source of modern ugliness—"the machine, for all its services to mankind, is so often represented in the role of Beast in its relation to Beauty in the life of the world that it has come to stand as a symbol of ugliness" (149, 28)—a shift in sentiment away from the feeling that designers were incapable of designing products for mass production that had any redeeming aesthetic qualities was underway. A general call was being heard for new forms of industrial-arts products that would be accepted as works of art, genuine in their own way, and as capable of raising the "standard of machine work within the scope of its possibilities." (139, 333)

Leon Volkmar, in the *New York Times,* called for a new form of art: "If you must have the machine, evolve a new type of beauty that will express the machine plus intelligent direction." (186) Volkmar took it as misguided sentiment to attempt to apply hand-originated decoration to machine products. It was important to him that machine-made products, whether they served luxury or convenience, should be acknowledged as respectable symbols of affluence. In this way those who would otherwise have had nothing, since they could in no way afford handmade originals, could now be proud of their material acquisitions no matter what their source or the quantity of their production. Richard F. Bach, however, defended the use of the machine to duplicate handmade historical styles in his comments to the Institute of Art and Science at Columbia University in 1922. It seemed entirely logical to him that modern technology should be used to replicate products and ornaments of the past.

Lewis Mumford was convinced that there was a fundamental difference between the aesthetic aspirations of handmade and machinemade products. It seemed to him that what was a virtue in one was anathema in the other. The craftsman, Mumford believed, was possessed by his work and dissatisfied with it until he had poured himself into every part. As a result, the value of his product was in direct relation to the amount of creative energy that had been consumed in its making. What Thorstein Veblen had termed conspicuous consumption was thus taken as the hallmark of handicraft aesthetics. The craftsman realized that convoluted rather than geometric forms were more logical for handmade forms and that, rather than mechanically pure surfaces, textured and ornamented surfaces could be relied upon to provide an acceptable finish to his work. On the other hand, good machine-made products, according to Mumford, demand "a complete calculation of consequences, embodied in a working drawing or design," and, as a result, "the

qualities exemplified in good machine-work follow naturally from the instruments: They are precision, economy, finish, geometric perfection." "A good pattern in terms of this mechanized industrialism," said Mumford, "is one that fulfills the bare essentials of an object; the 'chairishness' of a chair the 'washiness' of a basin; and any superfluity that may be added by way of ornament is in essence a perversion of the machine process, and by adding dull work to work that is already dull, it defeats the end for which machinery may legitimately play a part in humane society; namely, to produce a maximum amount of useful goods with a minimum amount of human effort." Mumford proposed that there was a canon of industrial aesthetics by which anything that demands more of a machine than it can logically or freely give "adds to the physical burden of existence." (156, 38) Once we refused to accept the norms of handicrafts as absolute, he suggested, there was a new kind of beauty to be achieved in machine-made products. Mumford was concerned about the fact that, although the logic of the machine was being acknowledged as stimulating a modern style, Americans had yet to allow for the "vagaries of human psychology" that would have to be taken into consideration if the style was to have real value. He was convinced that the machine was as "incapable of yielding fresh designs as a mummy is of begetting a family." (157, 263)

It is reasonable to believe that, once the freedom to modify an object in the course of its making that had been associated with the handicrafts was replaced by the preplanning and design that is essential for machine production, the forms of manufactured products would come under the control of design instruments (such as compasses and straightedges) and their production surrogates (lathes, mills, and planing and rolling machines). The action of instruments and machines was essentially geometric, as were the forms of the materials available for production (such as sheets, tubes, and wires). As a result, it was natural at the time to presume that geometric forms were most natural to machine production and that such qualities as mathematical accuracy and purity of finish would bring pleasure to those seeking manufactured perfection.

The new sensitivity to the aesthetic potential of utilitarian products and the recognition that machine production imposed unique design conditions were beginning to shape the character of industrial design. This character had found expression, unintentionally at first, in the work of those manufacturers who produced the unassuming vernacular products of everyday life. Its influence was echoed in the growing belief that everything had a rational base, and that the arts as well as social and industrial systems could be expected to operate smoothly and efficiently in order to bring culture and convenience to every citizen. It seemed reasonable to assume that the Americans, in particular, would welcome the new aesthetic of technology in their products and environmental arts. They anticipated a life of electrified ease in steel structures that scraped the sky. Mechanization was rapidly being transformed into an art compressing experience into moving pictures and expanding human capacity with speeding automobiles and airplanes. It seemed inevitable that the American cultural establishment would sense the groundswell of the modern movement and recognize a new kind of expression. But such was not to be the case.

Somehow, despite the overwhelming evidence of the growth of the modernist movement abroad, the American tastemakers looked the other way. As the defenders of democracy abroad in the recent war, perhaps they were inspired to honor their own origins by searching for what they believed to be a uniquely American style. In the process the searchingly original Arts and Crafts movement was aborted, Tiffany's inspired originality was forgotten, and Stickley's mission in furniture and furnishings design was transformed into the so-called Spanish Mission or Spanish Colonial and the English-based Colonial styles.

The Spanish Colonial style was the result of a deliberate effort by the newly founded (1920) Association of Decorative Arts and Industries to discover a "native" style and then to promote it into a commercial success as the "most intelligent substitute possible to find for the old, slow-growing development of a style." (179) The association proposed a truly New World character

reflecting the romantic history of some 300 years of Spanish and Indian traditions in the Southwest. In a major promotion to launch the Spanish style, the association exhibited in 1920 at the Grand Central Palace in New York a Spanish-American bungalow that was described as a "laboratory" of home furnishings. Designers, it was presumed, could now "create a new impetus toward the designing of fabrics, furniture, wall paper, etc., which shall be wholly American in spirit and, therefore, more fitting to the purpose of interior decoration for the homes of Americans." (111, 35)

The parallel movement in the Northeast in the early 1920s was devoted to rediscovering, preserving, and reconstructing the artifacts and even the environment of the American colonial era. Those who could afford to do so took it as their cultural and patriotic obligation to collect the modest folk arts and artisanal masterpieces of the country's first century and a half. Many of the major collections of Americana were started in the 1920s. The colonial scene was reborn when the Rockefellers rescued Williamsburg from oblivion. Dupont collected bits and pieces of the era at Winterthur, as did Henry Ford at Dearborn Village. The Vanderbilts created Sherbourne from scratch, just as others did at Sturbridge and Cooperstown. And the manufacturers of furnishings found no design challenge or risk in replicating the furnishings of the colonial era. The offices of the progressive J. Walter Thompson advertising agency in Manhattan's modern Graybar Building were transformed into a veritable colonial scene. It was, and still is, comfortably assumed that the Colonial style of the Northeast "reproduced the charm of Colonial simplicity without sacrificing the essential comforts and conveniences of modern life." (113, 48)

The Spanish Colonial style was a curious blend of Spanish baroque fashions with Indian forms and the Northeastern Colonial, an agglomeration of Jacobean, William and Mary, Queen Anne, Georgian, and rococo influences with traces of German folk art and French and Italian Renaissance, tossed into a cultural salad to nourish the Americans' transplanted spirit. Its best products were two styles of furnishings that have come to be considered natural to American domestic interiors.

The Spanish or South-
western style was arbi-
trarily conceived in a
deliberate search for an
indigenous American de-
sign expression. It traded
on the Arts and Crafts mission in furnishings and the
Rudolf Valentino brand
of romance. This room
blithely blends Spanish,
English, and Italian ideas.
Arts and Decoration, Jan-
uary 1925.

The Colonial style, considered "wholly American" in spirit, enjoyed a popular revival in the 1920s. This 1922 sketch from the Stanton Studios shows a hall and a living room. Art-in-Trades Club yearbook, 1922.

Joseph Urban's dining room in the Wiener Werkstatte style was as close to a modern expression as the Art-in-Trades annual exhibition ever came. Arts and Decoration, October 1923.

Two exhibitions staged by the Art-in-Trades Club at the Waldorf-Astoria hotel in 1923 and 1924 serve best to illustrate the state of mind in the early 1920s with respect to the industrial and decorative arts. Each of the some 50 rooms in the exhibitions was designed and decorated in a different historical style to emphasize the club's commitment to "beauty of form . . . in manufactured articles." (111, 33) The 1923 show included a Chippendale room (by Harry Wearne, the club's president), an English Renaissance or Tudor room, a Queen Anne reception room, a Louis XVI boudoir, and a Spanish Colonial patio and reception room. There was a token room in European modern—a dining room by Joseph Urban in the style of the Wiener Werkstatte. The 1924 displays revealed "the adaptation of historic decorative modes to modern usages" by including a Georgian library, an Elizabethan living room, a Spanish-Italian entrance hall, and, in particular, a Colonial breakfast room and apartment. One visitor to the exhibition somewhat cautiously noted that, although "one might wish for more that is new and unique, expressing the present, rather than finding its roots in the past," it was quite evident that historic styles still commanded the interest of most Americans. (112, 62)

In retrospect it is obvious that the American tastemakers and manufacturers preferred to remain fixed aesthetically to the past. A few advocates of the modern movement in the industrial arts warned that there was a desperate need for schools to train designers, but the majority believed that the wiser course was to continue the established practice of letting the Europeans pay for design education and then importing such samples and designers as were needed. America was "the greatest industrial art market in the world," declared the New York Times. Therefore, there would be a strong inducement for foreign talent to come to the United States. (183)

Those who controlled the cultural establishment were not entirely ignorant that a modern spirit in design was gathering momentum abroad. They had first sensed its coming in the decline of Beaux-Arts eclecticism and the bright but brief flare of Art Nouveau, then had watched the healthy thrust of the Arts and Crafts movement and the awakening of rationalism in design be diverted (at least momentarily) by the Decorative Arts movement, with its fawns, ferns, and fairies. They had sensed that the playful manipulation of natural forms was clear evidence that artisans were incapable of containing the impact of the coming machine age. However, it fell to the most sensitive talents of the time to capture the essence of the modernist movement first. Foreign artists like Picasso, Léger, Brancusi, and Mondrian, by rejecting maudlin sentimentality, found a way from the natural world into the dynamic sphere of abstraction and machinelike expression. Although the Americans had been warned publicly about their conservatism by the unexpected vitality of the Armory show a decade earlier, most still preferred to believe that the modernist movement was no more than a temporary aberration of aesthetic judgment that would be readily forgotten after the World War.

The French designers and decorators, however, realized that their artists had provided them with the substance of a new style. "To create, to build up a style," acknowledged André Marè, "the broad insight of the informed artist is needed, plus an infinite patience. . . . The same fire and faith must animate the artistic evolution of a chair no less than that of a painting or a sculpture." Marè, with his partner Louis Sue and others like Paul Follot, André Groult, Maurice Dufrène, Mallet-Stevens, Pierre Legrain, Emile Jacques Rhulmann, and Paul Poiret, found a common foundation in geometricity and helped to build it into a homogeneous style in the modern spirit, which proposed that the pristine qualities of geometric forms were more innocent and therefore more honest. Marè found justification for the modernist style in examples from the past: "When we study masterpieces of art and architecture we can see that their harmony is confined to geometry. Certain lines and proportions have peculiar virtue; but not all triangles, squares and ovals are equally beautiful. So, after reflection, we adopt some as the basis for our designs. . . ." (109, 331)

In 1922, Leo Randole attempted to explain the modernist movement to his American colleagues: "It is the Gallic soul imprisoned within geometry and informed with the restraint imposed by the intellect through which beauty triumphs in her own way. In a nation where the gift of art is as traditional as it is innate, even the most revolutionary modernist turns constantly to the past for his inspiration, his justification. The modern decorator in his intellectual asceticism penetrates the past through the logic of geometry, which so infallibly defines the equilibrium of all harmony." (109, 331) American manufacturers of decorative products, however, remained particularly insensitive to the growing influence of modernism. They were preoccupied with their adaptations of traditional styles, and unwilling to be diverted by threats of a new style—at least until it had become popular enough to make copying profitable.

The French designers found their own manufacturers more receptive. Industries adjusted their manufacturing technologies to suit the radical ideas, thus breaking with their dependence upon established methods and entrenched aesthetics. Furthermore, the modernists took it as their mission to "democratize" modern art, to bring it within the reach of all classes in order to satisfy their craving for the beautiful." In the process a new collaboration developed between designer and manufacturer that resulted in one of the basic contractual methods now familiar to industrial designers. It became customary in France for the manufacturer not only to pay the designer for his prototype expenses, but also to reimburse him with a royalty for every reproduction of his design. The concept of royalty thus obligated the designer to share in the risk and problems that might be encountered by the manufacturer. The understanding that developed between conceiver and producer was of great benefit in the refinement of aesthetic concepts for machine production. "The mating of creative minds," it was acknowledged at the time, "imbued with the aesthetic impulse, to the commercial skill which makes possible the wide distribution of beautiful objects is a characteristic of the hour in which we may all rejoice." (109, 370)

Paul Poiret was convinced that "new forms exist only in objects which are created for our modern needs, and since nowadays the need of comfort and mechanics dominates everything else, it is only in objects in which art applies itself logically and intelligently to mechanical laws that real beauty and new forms can be found." "In years to come," said Poiret, "scientists will probably be able to explain the force that prompts the instinct of the artist to master the equilibrium of masses and volumes in perfect unison with the machine." (104, 380) Thus Poiret provided the modernist cause with its basic creed: that art may be used intelligently and rationally in order to achieve harmony between human needs and manufacturing capability.

Though not as many French designers were working in the new style as one might imagine, they had a strong influence. Their geometric style took hold first at the highest fashion level, where its daring forms were easily adapted to decorative patterns for clothing and accessories, to cosmetic packaging, and to the related advertising. One finds a clear connection between the decorative glass of Lalique and Baccarat and the bottles and packages of Francois Coty and other French perfumers, and it was only a short step from the arts of personal adornment to those of the environment. The modernist style began to appear in fabrics, furniture, lamps, and other accessories. Art directors found fame and profit by combining utility and beauty in the modern mode to make their clients' products attractive to an ever-growing public.

French prestige in the arts of "haute couture" in the United States was given added impetus by an exhibition held in 1924 at the Grand Central Palace that consisted of a series of *salons intimes,* one for each of the French perfume makers. The most successful exhibit turned out to be the full-scale replica of Coty's Parisian salesroom at the corner of Place de Vendome and the Rue de la Paix. Thus, the modernist style was familiar to at least one social level in America before the 1925 Paris exposition. In fact, Paul Poiret had already promoted Art Moderne in a successful lecture tour of the United States in 1923 (after which he had announced that he was discouraged to find extensive pirating of his designs by American wholesalers). On the other hand, the few Americans who tried to market their own modernist designs found that American manufacturers were still reluctant to devote any capital to native ideas. Nevertheless, as the mass production of comparatively inexpensive clothing and fashionable accessories was becoming highly organized in the United States, a democratized market for style was beginning to gather momentum in which women did not object to wearing almost identical outfits—providing they were in the latest Parisian fashion.

In an effort to capitalize on the rekindled interest in French styles, France determined to regain her prewar eminence by rescheduling for 1925 the "Exposition International des Arts Décoratifs et Industriels" that had originally been discussed in 1912 and planned for 1915. "Will we offer to humanity the birth of a new French genius, or will we, ourselves, receive and follow the formula of others?" wrote Edmond Haraucourt, director of the Musée de Cluny in his 1921 call to his countrymen to join the intellectual and artistic communities in the great exposition. (108, 91) However, in deference to the increasing democratization of design and the growing importance of mass production, the planners of the exposition decided to shift the emphasis from the one-of-a-kind masterpieces that had characterized earlier international expositions to the objects of everyday life, and to reward originality rather than subservience to historical motifs. The French were confident that their creative vigor would stand out above that of other countries: "Let us export our intelligence, our elegance, our art. In the markets of the world let us be the merchants of beauty. . . . a civilization characterized by the abolition of time and distance, by the reign of electricity and mechanical invention requires art to its liking." (187) Logic, truth, and harmony were proposed along with science as fundamental to contemporary art. By the time the Paris exposition opened, the conventionalized sentimentality of stylized animals, sexless nudes, and stone roses had been invaded by the modernist geometric forms of the machine-age style. Today

This interior, with furniture
and shelving in polished
metal by Mallet-Stevens
of Paris, displays innovative yet consistent use
of rectilinear geometry
as a basis for design.
J. Prouve, Le Métal
no. 9.

Pierre Legrain's automatic
piano for Auto-Pleyela, not
considered outstanding
in its time, persists as a
valid attempt to turn a
functional product into
a piece of abstract tech-
nological sculpture.
J. Prouve, Le Métal
no. 9.

René Lalique's glass
sculpture and containers
combined the votive
nudes of the era with ab-
stractly modern forms. The
powder-puff container by
Coty is one of the clas-
sic cosmetic packages.
M. Battersby, The Dec-
orative Twenties.

the two styles have been, somehow, melded into one—Art Deco—that conveniently uses the abbreviated title of the first to define the forms of the second.

When the directors of the Paris exposition sent out invitations to participate, every country except Germany received one and all but two countries accepted the challenge. China was not in a position to accept because of its chaotic political condition, and the United States declined because Herbert Hoover, as secretary of commerce in 1923, had concluded that the Americans could not comply with the exposition's requirement that only original work could be exhibited. Subsequently, Hoover turned the French invitation over to the Smithsonian for consideration, but again because of the lack of support from commercial interests the State Department did not recommend appropriations. Later, in 1925, Hoover explained his action in a telephone address to the 4th Annual Exposition of Women's Arts and Industries:

When the United States was invited to participate in this exposition, I canvassed the various interested manufacturers to learn if they thought it advisable for our government to undertake an American building in Paris. The advice which I received from our manufacturers was that while we produced a vast volume of goods of much artistic value, they did not consider that we could contribute sufficiently varied design of unique character or of special expression in American artistry to warrant such a participation. Whether this be the actual case or not, it was their opinion. My own conclusion from this experience was that it was high time that we should begin to develop more design and artistry. Therefore, as Secretary of Commerce I requested a number of men and women prominent in the field of industrial arts in this country to constitute themselves a delegation to visit the Paris Exposition. I am hopeful that this mission may serve to promote the growth of our American industrial arts. I know of no field where women can be of greater service in such design, and in the stimulation of the more artistic presentment of articles of common utility. Art in its more extreme form cannot find a place in every home. We cannot all of us afford galleries of paintings or rare sculpture. But we can see to it that our homes and the furnishings that go to make up those homes shall be in accordance with canons of art and taste. (214)

The refusal of the United States to join its allies and other nations in the first great postwar fair, the Paris Exposition of Modern Decorative and Industrial Arts, shocked the cultural establishment, even though in retrospect this establishment may be held responsible for the fact that America had nothing original to show. And, though Hoover's decision did not surprise the manufacturing and merchandising industries, the Americans were galvanized into demanding prompt government action—largely, one might suppose, because of the blow to U.S. prestige and the economic threat that was posed by the exposition. As a result, Hoover was obliged to honor a demand from Charles Richards that he delegate a special commission of 100 representatives from interested trade associations, and a few token architects and designers, to visit and report on the exposition for the benefit of American trade. Other American designers, such as Russel Wright, Donald Deskey, and Walter Dorwin Teague, were drawn to Paris and returned as champions of modernist art.

The report of the American commission was self-serving yet apologetic:

As a nation we now live artistically largely on warmed-over dishes. In a number of lines of manufacture we are little more than producing antiquarians. We copy, modify and adapt the older styles with few suggestions of a new idea. It is true that this practice of reproducing the older forms has been an invaluable education to our people. It is also true that the adaptation of old motives when performed with intelligence and skill continues and probably will continue to give us a large proportion of the decorative manifestations acceptable to American taste. It would seem equally true, on the other hand, that the richness and complexity of American life call for excursions into new fields that may yield not only innovations but examples well suited to the living conditions of our times. (73, 22)

Americans needed to make room for new ideas, the report continued, because they would not avoid the modern movement in the applied arts; therefore, since the movement would reach the United States shortly, Americans should "initiate a parallel effort of our own upon lines calculated to appeal to the American consumer."

However, the members of the commission did not all agree on the importance of the modern movement. Richardson Wright, the editor of *House and Garden,* expressed doubt that the modern movement as applied in the home would ever get a strong foothold in the United States. He saw nothing unusual in the fact that there were not enough people in the country who were interested in the new style to justify an American exhibit in Paris. The opposite view was expressed by Leon V. Solon, who wrote in the *Architectural Record* that the commission's report represented a change of heart on the part of industry with respect to design and suggested that the new styles did not "proceed from an altruistic impulse, but from one which is infinitely more solid and significant." "This sudden interest in decorative excellence," wrote Solon, "is a policy of expediency, compelled by the extraordinary improvement in public taste and the economic value of artistically treated goods. . . . It is the industrialist who now takes the risk in innovation, not the solitary individual laboring in comparative obscurity." (199, 181)

The commission did not seem to recognize that the Europeans had deeper resources from which new ideas could be drawn, that their governments and industries respected and supported innovative design training, and that they saw new concepts in form as signs of creative vigor and not as threats to the cultural status quo as they appeared to be in the United States. Nevertheless, the trauma of Hoover's rejection, plus the phenomenal success of the Paris exposition, swept away much of the American resistance to the modern movement. Within a year Art Moderne, or modernist art and decoration, was being eagerly duplicated by Americans. Lewis Mumford caustically noted that nothing had changed: "American designers, instead of designing directly for our needs and tastes, are now prepared to copy French modernism, if it becomes fashionable, just as they habitually copy antiques." (155, 576)

The Paris exposition became a dramatic showcase for the talent and creative imagination that had been constrained by the austerity of the recent war. It provided a great impetus to the modernist movement in the form of a series of pavilions whose interiors had been subsidized by the French minister of fine arts to be designed and built by the major Parisian merchandisers. Department stores such as the Galeries Lafayette, Bon Marché, and Printemps and the Magazins du Louvre gave their staff designers a free hand to conceive and produce a series of rooms in the modernist style. They were called "ensembles," in recognition that every item in each room was to be unique and designed to be in aesthetic harmony with all of the others. This provided a homogeneity of expression that is typical of historical styles, and thus helped to make the modernist style comprehensible and palatable to the general public. With their bold geometry in combination with the more sentimental decorative arts and the bright flashes of Bakst's "Russian Ballet" colors and patterns, the rooms were interpreted as complete happenings — exclusive creations, rather than prototypes of the democratic expression that was claimed for them. Some "ensembles" were purchased in their entirety as unique works of art; others were picked apart for inspiration by visiting designers and manufacturers.

While the pavilions were competing for attention, off in a corner of the exposition grounds stood a small house built by Le Corbusier and his followers, under the banner of "L'Esprit Nouveau," in the clean machine-age style that would eventually displace Art Deco and other short-lived fashions to become the dominant style of this century. Corbusier's pavilion was clearly in line with his convictions that "the new dwelling house must be a machine for habitation" (as quoted by Mumford) and that it and its furnishings should be standardized for machine production and be replaceable rather than permanent. (158, 78) In this respect, French architects and designers were not entirely out of step with the philosophy of design that was taking shape at the Bauhaus at Dessau. Although there was as yet little rapport between the recent enemies, by 1923 Le Corbusier was printing in *L'Esprit Nouveau* a sympathetic commentary of the work of the Bauhaus and a paper from Walter Gropius asking for support.

The cover of the 1925 Paris Exposition catalog clearly illustrates its original focus on the decorative arts rather than the Art Moderne for which it is best remembered. Report on the International Exhibition, Paris 1925.

The Paris Exposition's break with the past is reflected in the comment by Helen Appleton Read that passing through these obelisklike silver towers of the Porte d'Honneur was not unlike coming upon a cubist dream or a city on Mars. Report on the International Exhibition, Paris 1925.

The main entrance to the Paris Exposition, the Porte de la Concorde—towering pylons and a bronze statue of "Welcome." Report on the International Exhibition, Paris 1925.

MINISTÈRE DU COMMERCE ET DE L'INDUSTRIE

PARIS-1925

EXPOSITION INTERNATIONALE DES ARTS DÉCORATIFS ET INDUSTRIELS MODERNES AVRIL-OCTOBRE

Pierre Patout's handsome "Pavilion for a Rich Collector," created for the celebrated French decorator Emile-Jacques Ruhlmann, was built of reinforced concrete and was honored as a harbinger of the future for its cleanliness, its beauty, and its labor-saving approach to structure. Report on the International Exhibition, Paris 1925.

Janniot's group of stone figures outside the Ruhlmann pavilion combined nudes, animals, and flowers in forms that had clearly broken with the past; however, the spirit was classical. Report on the International Exhibition, Paris 1925.

The salon of the "Pavilion for a Rich Collector," by Jacques Ruhlmann. M. Battersby, The Decorative Twenties.

Although it was hidden away in a corner of the Paris exposition, the small pavilion of L'Esprit Nouveau, *by Le Corbusier and Pierre Jeanneret, was a landmark. Constructed entirely of standardized elements, it was a prelude to manufactured dwellings.* L'Esprit Nouveau, 1925.

The quietly classical interior of the L'Esprit Nouveau *pavilion may be taken as a forerunner of the modern style. Its spare monumentality necessitated the visual relief of the paintings, which were by Leger and Jeanneret.* L'Esprit Nouveau, 1925.

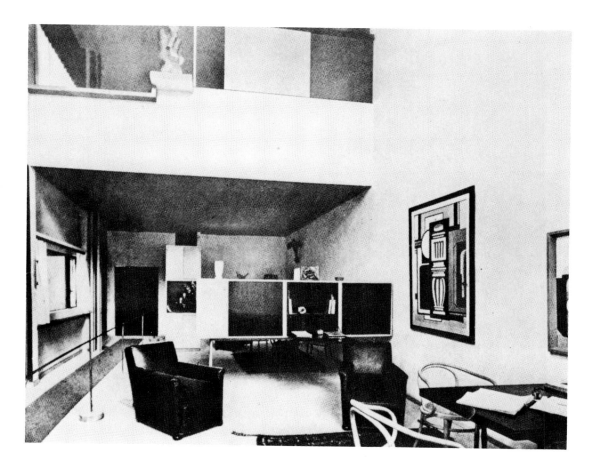

Le Corbusier's terrace for a villa at Ville d'Avray illustrates the International Style's debt to the great ocean liners of the day. J. Prouve, Le Métal no. 9.

The Swedish exhibits at Paris were characterized by a restrained sense of design and a simplicity of form, as in this interior by David Olmberg. Report on the International Exhibition, Paris 1925.

This engraved crystal covered bowl and platter, designed by Simon Gate and made by the Orrefors Glass Works of Sweden, shows the new Stockholm town hall surrounded by classical yet contemporary ornamentation. Report on the International Exhibition, Paris 1925.

It is conventional to credit the new style to the Germans, but the fact is that, for the most part, design in Germany and France was moving along parallel paths, with the Germans committed to the principle of design through craftsmanship and the French convinced that expression through geometry would produce forms that were the true expression of the machine age.

The Swedish pavilion became the center of attention in Paris. The astonishing number of awards won by Sweden, whose Arts and Crafts movement has evolved without a break from the rural folk arts into an urbane aesthetic—35 grand prizes, 46 gold medals, and many other lesser awards—provided the momentum that gave Swedish and other Nordic designers preeminence in the decorative arts for decades to come. To a large extent their success was due to the fact that the Swedish Society of Arts and Crafts had determined that industrial as well as hand-made objects must speak the language of the period to which they belong. The result was a freshening of attitude toward design and the development of forms that suggested a new clarity in the decorative arts that was entirely Scandinavian in spirit and vitality.

The Swedish Society of Arts and Crafts decided, after a long history of Swedish crafts (the society was founded in 1845), to challenge tradition and, through the leadership of Gregor Paulsson, director of the society, and Dr. Eric Wettergren, curator of the Swedish National Museum, became insistent on "the need for new art forms to suit our modern age." (193, 79) The philosophy was directed against the use of machine production to imitate the handicrafts by producing artificial and cheap reproductions of costly objects, and encouraged Swedish handicrafts to develop forms that reflected the precise and pristine forms that one associates with machine-made products. As a result, the modern Swedish handicrafts found a handsome accommodation between the simpler forms of industry and the spare yet rich details that were most logical to the virtuosity of the craftsman. The United States, on the other hand, had not only rejected the modern movement in design but had also denied

earlier nourishment to the flowering Arts and Crafts movement before it could come to the level of maturity that it reached in Sweden.

Art Moderne Becomes
Industrial Design

We passed from the hand to the machine, we enjoyed our era of the triumph of the machine, we acquired wealth, and with wealth education, travel, sophistication, a sense of beauty; and then we began to miss something in our cheap but ugly products. Efficiency was not enough. The machine did not satisfy the soul. Man could not live by bread alone. And thus it came about that beauty, or what one conceived as beauty, became a factor in the production and marketing of goods.

Ernest Elmo Calkins, 1927 (121, 147)

One of the results of the Paris exposition was to awaken the public press in the United States to the existence, if not the importance, of modern design. Whereas the specialized magazines in art and decoration had been supporting the strong interest of the establishment in traditional styles and the cultural aspects of design, now American newspapers and magazines began to promote design for its economic value. It was stated that America was beginning to catch up as big business was losing its "timidity in the presence of Art." An editorial in the *New York Times* challenged industry: "American captains of industry must soon change their traditional characters. True, the public may still cling to its conception of them as iron-jawed, level-headed giants of efficiency. . . . But if the present international competition keeps up, we may find them, like the Germans, openly cooperating with museums and industrial art schools; or, like the French, proclaiming themselves merchants of beauty." (187)

The various American decorative-arts associations were quick to point out that the French had only achieved what they themselves had been proposing all along: the collaboration of artists and designers with industry that would result in better and more marketable products. The Art-in-Trades Club issued a manifesto in 1926 challenging all American artists to compete in a competition for modern furnishings. Others, however, recalled that as early as 1923 it had been proposed that American manufacturers did not need to produce cubistic chairs or futuristic consoles, "because our native sense of the fitness

of things and our demand for the beautiful serviceable article, as opposed to the irritating, eccentric one, comes to our rescue in furnishing our homes and business offices." (110, 18) It was even suggested naively that, since the modernist style belonged to the Europeans, Americans should look to their own native sources and draw inspiration from the Alaskan, Mayan, and Toltec cultures.

American museums moved quickly to capitalize on the interest in design and the decorative arts that had been sparked by the success of the Paris exposition and to reaffirm their hope to be tastemakers for the public. Within a year after the close of the show in Paris the American Association of Museums had organized and imported a selection of products from Paris, primarily French and Swedish, to tour its member institutions, beginning in February 1926 at the prestigious Metropolitan Museum of Art in New York. The preface of the catalog for the show illustrated once more the museums' well-meaning but short-sighted attitude toward American design: "The collection has been brought to America not to stimulate the demand for European products nor to encourage copying of European creations, but to bring about an understanding of this important modern movement in design in the hope that a parallel movement may be initiated in our country . . . to bring to us new forms appropriate to the living conditions of the twentieth century." (216)

A second stage in the new role of the museum as a cultural reporter was reached in 1927 when the Metropolitan Museum provided space for the first time to a foreign country in the field of the decorative arts. In a move that inaugurated the reign of Scandinavian design in the United States, the museum staged a special exhibition of Swedish industrial arts. The net effect of the exhibition was to demonstrate a new and sincere attempt to speak the language of the day—a language of simplicity and truth. Some cultural historians have pointed out that the Scandinavians had long been familiar with the spare elegance and reverence of the Shakers' furniture and their other ingenious and appropriate objects

A Selected Collection of Objects from the International Exposition of Modern Decorative & Industrial Art

P A R I S

1 9 2 5

Organized and Exhibited by
The American Association of Museums

The American Association of Museums was quick to arrange for a national tour of objects selected from the Paris exposition, and borrowed the art from the original catalog.

for everyday living. Perhaps the appeal of Scandinavian design to Americans stems from a renewed conscience for design that has sent hundreds of American craftsmen and designers to study and work in the studios of Scandinavian designers and has made titles like Swedish Modern and Danish Modern part of the American design vocabulary.

Once American museums had bestowed cultural respectability upon the modernist movement, it became inevitable that department stores on this side of the Atlantic would emulate the Parisian stores at the exposition by staging their own exhibits and supporting the design and production of products in the new style. Even if the products displayed were too modern for the average consumer, they would at the very least attract the public into the stores and sharpen its appetite for other products. The French had displayed such rooms in the United States earlier, but not until the smashing success in Paris did the idea catch on. Now Macy, Altman, Wanamaker, Lord and Taylor, Franklin Simon, Marshall Field, and others decided to stage their own "ensemble" exhibitions in the style of the French. To this end it seemed natural that they should turn to the worlds of the theater and promotion for designers to help them conceive and install their exhibits, to decorate their show windows, to design some of the furnishings, and to offer other promotional ideas that would stimulate public interest. Lee Simonson, Norman Bel Geddes, Henry Dreyfuss, Russel Wright, and others were already well established as designers of stage settings, theatrical spectaculars, costumes, and decorative accessories before they were attracted by the glamor and excitement of the Art Moderne spirit to try their hand at staging the public debut of decorative and utilitarian products. It was a relatively small step for them to shift their attention from the make-believe world of the theater to the world of real products presented in make-believe environments.

In 1927, the R. H. Macy Company staged an experimental exposition to show the advances American manufacturers had made in introducing good design to everyday products. The public response was good enough to convince Macy to stage, in 1928, the most ambitious project of all of the department stores, an "International Exposition of Art in Industry." This show was a major activity of Macy's new department of design under the direction of Austin Purvis of Philadelphia. Lee Simonson, designer of sets for the Theater Guild, with Virginia Hamill as co-director, designed the background of the galleries and show windows for the exhibition, which consisted of an arcade of 15 rooms in the "ensemble" manner. The rooms themselves were designed by a selection of internationally known designers. Josef Hoffmann, director of the Academy of Industrial Art, designed a boudoir and powder room that included products from the Wiener Werkstatte. For France, Maurice Dufrêne designed a dining room and Joubert et Petit a studio living room. Bruno Paul, director of the German state schools of fine and applied arts, combined handmade and machine-made products in his designs for a man's study and a dining room. For Italy, Gio Ponti designed a country living room and, of all things, a butcher shop. The Swedish exhibits included Orrefors and Kosta glass in addition to metalwork, textiles, and ceramics. The American rooms were particularly interesting because they included not only handmade and manufactured decorative arts but also a selection of manufactured appliances. A three-room apartment by Kem Weber of California included a General Electric refrigerator, a Hotpoint electric range, and a sink and bathroom fixtures by Crane. William Lescaze's penthouse studio contained a sunlamp and electrical fixtures, and the living room by Eugene Schoen included lamp fixtures designed by him and handmade silverware by Peter Muller-Munk, the recent German émigré who was later to become a prominent American industrial designer.

The Macy exposition was the first useful opportunity for American designers and manufacturers to show their work in the modern spirit. Even so, the magazine *Architectural Record* still regretted the failure of the Americans, despite the pleas of architects like Frank Lloyd Wright and Louis Sullivan and designer-craftsmen like Gustav Stickley and Louis Tiffany, to find a connection between art and industry. It further expressed a sense of embarrassment that Americans had to go abroad to learn what America's own modern designers had taught the Europeans. Nevertheless, the demand for more handsome manufactured products had reached a point where it could no longer be denied.

Robert W. de Forest, president of the Metropolitan Museum of Art, in a museum monograph on the Macy exposition, speculated (as did many others at the time) on the most appropriate title for the emerging profession. He felt that although *industrial art* was not quite appropriate, it might have to be used. The German equivalent, *kleine Kunste* (small arts), which was used in reference to German industrial arts, carried, in de Forest's opinion, implications of inferiority, as did *applied art.* Although de Forest preferred to adopt the French phrase *decorative and industrial art,* he recognized that as practice expanded to include the design of all manufactured products—not only those that were decorative accessories— a broader and more serious title would be mandated. It was evident that a new title had to be adopted that contained the proper connotations but that had not been tied to another particular occupation. It remained for Ernest Elmo Calkins, head of the very successful Calkins-Holden advertising agency, to catch the character of the crystallizing profession in his seminal article "Beauty the New Business Tool" in the August 1927 issue of the *Atlantic Monthly,* in which he proposed that manufacturers now concern themselves with good design in their products. It became inevitable that *industrial design,* which had been used on and off for a decade, would emerge as the name of the new profession. Since then the appropriateness of the name has been questioned often because of its ambiguity. However, as it has achieved national credibility and has been adopted internationally as the generic name for the practice of the imaginative development of manufactured products and product systems that serve the physical needs and satisfy the psychological desires of people, the pressure for a different title has abated. Industrial design today is concerned with utility and safety as well as the meaning and beauty of the products that designers create for their fellow humans.

Lee Simonson's fifteen rooms at Macy's Exposition of Art in Industry were grouped around a series of show windows, such as this one, that displayed the best that each country had to offer. It was said that in this exposition the Americans looked to Europe to learn the lessons that America's pioneer modernists had taught Europe. Architectural Record, *August 1928.*

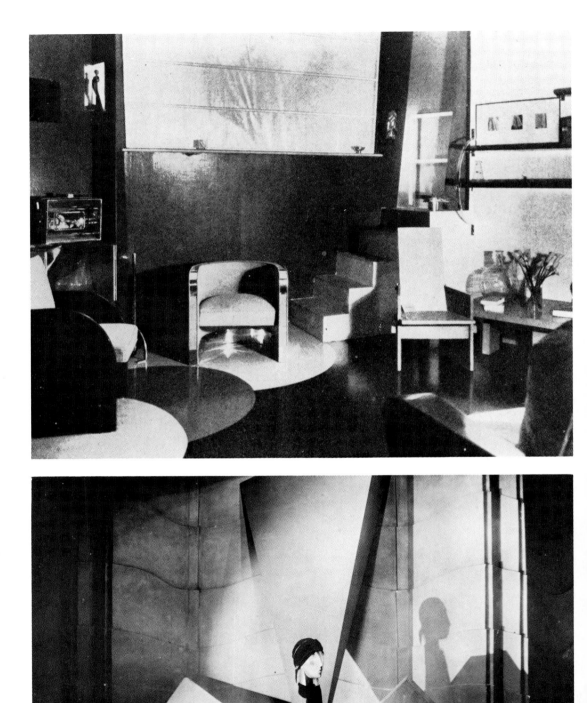

William Lescaze's penthouse for the Macy exposition displayed the then-fashionable affection for geometric forms and modern metals and composition materials, but lacked the stark spaces and arrangements of the International Style or the contrived modesty of today's "high tech" style. Architectural Record, August 1928.

This 1938 display window by Norman Bel Geddes for Franklin Simon combined three pieces of merchandise—a turban, a scarf, and a bag—and dramatic lighting. "The window is a stage," he said, "and the merchandise the players." Hoblitzelle Theatre Arts Library, University of Texas, Austin.

By 1928 the modernist movement was in full swing in New York as merchandisers vied with each other for public attention. The B. Altman department store held an exhibition of what it called "Twentieth Century Taste," with complete rooms in the modern style by American and French designers. Lord and Taylor announced an international competition in design and trade symbols. And Franklin Simon hired Norman Bel Geddes to develop a modern approach to its window displays. When the windows were unveiled the crush of viewers was so great that police had to be assigned to keep traffic moving. Within three months every Fifth Avenue store had shifted its displays to the new style.

In a letter to the *New York Times,* James Rennie criticized Bel Geddes for having abandoned the theater for industrial design. "Norman Bel Geddes has left the theatre flat," Rennie wrote, "and is now designing window displays, automobiles, scales and various other odds and ends which have no relation to dramatic art." (188) Bel Geddes responded that industrial objects offered broader opportunities than the theater in its current condition:

The theater is a fickle mistress. We live in an age of industry and business. Industry is the driving force of this age and art in coming generations will have less to do with frames, pedestals, museums, books and concert halls and more to do with people and their life. It is the dominating spirit of this age. . . . It is as absurd to condemn an artist of today for applying his ability to industry as it is to condemn Phidias, Giotto or Michaelangelo for applying theirs to religion. It is more important that I should be working at something that interests me, that is of the present, such as the automobile, the airplane, the steamship, the railway car, architecture and furniture, than it is for me to keep working in the theater merely because I spent 15 years doing so. (189)

As one might expect, paternity for the growing profession was claimed by its earliest practitioners. Norman Bel Geddes was convinced that he had established the profession in 1927, when he had been the first designer of national reputation to surround himself with a staff of specialists in order to offer industrial design services. One of his first assignments had come in that year from Raymond Paige of the Graham Paige

Raymond Loewy's first major industrial design assignment was the redesign of the Gestetner duplicating machine in 1929. Raymond Loewy.

Motor Company, who had asked Bel Geddes to conceive ideas for an automobile five years in the future. Raymond Loewy claimed that "the beginning of industrial design as a legitimate profession" had been his contract with the Hupp Motor Company because "for the first time a large corporation accepted the idea of getting outside advice in the development of their products." (60, 85) The first product of his design that was manufactured, however, was the Gestetner Duplicator in 1929. When Walter Dorwin Teague had come back from Europe in 1926 determined to shift his work from graphics to the design of industrial products, he had added "industrial design" to his letterhead. When in 1928 he was recommended by Richard Bach of the Metropolitan to Adolf Stuber, son of the president of Eastman Kodak, to design the interior of his home, Teague took Stuber on a tour of the modern art shops in New York, including the contemporary shop of Rena Rosenthal, in order to show him the clean lines of home furnishings that were coming into style. From this experience their acquaintance grew into a lifetime contract for Teague as a consultant designer for Kodak. Joseph Sinel also believed that he should be given credit as the legitimate father of the profession because he was working on products for industry early in the 1920s through his commercial art work for advertising agencies like Calkins and Holden, McCann-Erickson, and Lennen and Mitchell. Among his first products for American manufacture were the Acousticon Hearing Aid and the International Ticket Scale.

Others were also designing products for industry by 1927. Donald Deskey was preoccupied with a candy vending machine, Egmont Arens was designing and seeing to the manufacture of a line of lamps, and George Sakier was serving as design director for the American Radiator and Standard Sanitary corporations.

In point of fact, none of these early practitioners invented industrial design—nor was it created by Henry Dreyfuss, John Vassos, Lurelle Guild, Russel Wright, Ben Nash, Thomas Neville, Francesco Collura, George Cushing, Harley Earl, William O'Neil, Gustav Jensen, Scott Wilson, Ray Patton, George Switzer, Harold Van Doren, John

Alcott, or any of the others who began to design products for industrial manufacture in the late 1920s. Nor was the profession conceived abroad and introduced to the Americans by foreign designers and academicians. Instead, industrial design emerged in the United States as a distinct calling in direct response to the unique demand of the maturing twentieth-century machine age for individuals who were qualified by intellect, talent, and sensitivity to give viable form to mass-produced objects. The new calling appeared almost simultaneously at many points in order to fill the vacuum left by the inability of craftsmen to anticipate every demand that would be imposed on a manufactured product by an expanding technology and a merchandising commitment to consumer satisfaction. These pioneering industrial designers and others like them were the results of this phenomenon, not its originators.

Perhaps no other single incident better illustrated that design had an important role to play than the competition for the consumer's dollar between the Ford Motor Company and General Motors in the mid-1920s. Henry Ford's Model T, first produced in 1908, had already entered folk history. Driven by millions, it was beloved as a means of basic transportation as much as it was criticized for refusing to keep up with the times in appearance. Henry Ford stood imperiously above the argument, autocratic and adamant in refusing every recommendation that he update the style of his product. That his homely contraption was selling over a million units a year, whereas its nearest competitor, General Motors' Chevrolet, had only a third as many sales, was good enough for him. Then, in 1923, when Alfred P. Sloan became president of General Motors, he determined to catch Ford in sales. In 1926 General Motors took a bold step in styling by introducing an all-new, snappy, and colorful Chevrolet in direct competition with the Model T. That year Chevrolet sold more automobiles than Ford. Henry got the message and finally retired the Model T. Within a year Ford introduced the more stylish Model A, launching annual model changes and deliberate styling as a marketing tactic. Henry Ford now found it convenient to reverse his opinion about art: "Design will take

more advantage of the power of the machine to go beyond what the hand can do and will give us a whole new art." (19, 28) By 1927 General Motors had already established a styling section, with Harley Earl in charge, and other major American companies were beginning to follow suit.

Thus, 1927 may be taken as the seminal year in the recognition of industrial design as an indispensable factor in manufactured products. The American design, merchandising, and industrial establishments had been jolted out of their reverie by the success of the Paris exposition and the subsequent popularity of the modernist products from France that were imported by museums and department stores. However, whereas in Europe the new style was primarily effective in the realm of the decorative arts and domestic furnishings, in the United States its influence was felt much more strongly in the areas of machinery, appliances, and vehicles. The major reason, no doubt, is that advertising agencies (especially those that served the major industries) recognized the promotional value of the more modern and better-looking products and urged their clients to employ industrial designers.

Ernest Elmo Calkins was one of the most enthusiastic supporters of industrial design. At one time or another he had known and employed as illustrators Joseph Sinel, Norman Bel Geddes, Walter Dorwin Teague, and others who went on at his urging to become industrial designers. Calkins credited to design the success of Chevrolet over Ford, and proposed in the article mentioned above that "the only art that can survive and grow is art that is related to our life and our needs, and that has a sound economic foundation." "It is to be hoped," wrote Calkins, "that manufacturers in the search for design to beautify their products will start with a clear conception of what beauty is, especially beauty in an article of use. . . . The surest guide in divining new beauty in machine-made things is to grasp and interpret the beauty they naturally and intrinsically possess." (121, 154) Calkins's article was acknowledged by others as providing a useful accommodation between the relative roles of the engineer and the designer in product devel-

The classic 1923 Ford
Model T four-door sedan.
Library of Congress.

This advertisement for the
1924 standard Ford coupe
suggested that the busi-
nesswoman of the day
had not only her own
office and telephone, but
also her own wheels wait-
ing at the curb. Vanity Fair,
May 24, 1924.

The "Fordor" Sedan of
1927, the last of the Model
T Fords—gaunt, direct,
and essentially a nine-
teenth-century product.
Courtesy of Ford Archives,
Henry Ford Museum,
Dearborn, Michigan.

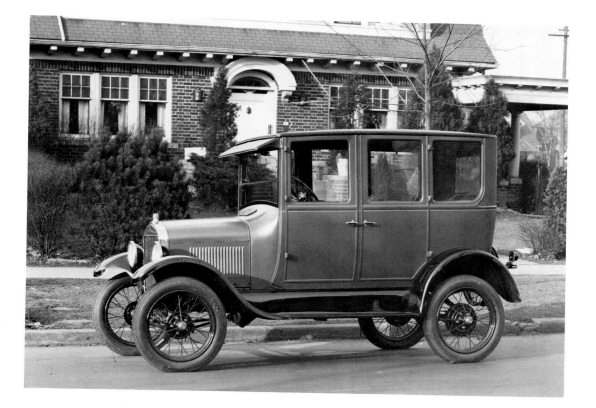

General Motors' Chevrolet turned the tables on Ford's dominance of the low-cost market, proving that form and appearance had power in the marketplace. Automobile Manufacturers' Association.

Henry Ford caught on fast. Within a year after he stopped Model T production, the more stylish Model A was on the market and Ford was back in the race for first place in sales. Automobile Manufacturers' Association.

opment by suggesting that, once technical and manufacturing commitments have been met, the appearance of the product becomes of paramount interest to the consumer. An article in the *Review of Reviews* in 1928 gave a great deal of credit to advertising men like Calkins as the first business group to appreciate the cash value of advertising products that were good-looking. And a letter by N. Shidle that was published in *Automotive Industry* on August 13, 1927, acknowledged that "from now on, the pulling power of beauty having been so firmly established, it is expected that beautification may follow on the heels of mechanical improvement more rapidly than in the past." (198, 219)

Richard F. Bach was convinced by 1928 that his efforts through the Metropolitan series of annual exhibitions of the decorative arts to awaken Americans to the value of design had finally been rewarded. "Style," he wrote, "is at last coming to recognition once more for what it truly is: the real drawing power in a thousand products, the basic selling point and—given a good design to begin with—the most convincing argument for quality." (115, 599) It had become apparent to Bach that the stream of design vitality had altered its direction and that the commercial world had moved closer to the modern styles from abroad. As a result he decided that the eleventh exhibition in the museum's series should adopt the French pattern of room "ensembles" and be used to demonstrate that a style of design could generate ideas without an "exuberance of novelty and with never too strong a regard for sales value." (204, 23) The show, which opened in February 1929, was called "Contemporary American Design." Although it was intended to run only through March, by the end of the period 100,000 people had seen it and the Metropolitan found such a great demand that it was obliged to extend it to October. This exhibition neatly brackets a period of transition in American design from subservience to traditional styles to an awareness, if not a total acceptance, of the modern. The subtitle of the exhibition, however, was "The Architect and the Industrial Arts," and it bypassed the handful of architects and designers who had pioneered the new spirit of design in the United States and the growing band of talented industrial and interior designers, electing instead to feature rooms designed by established American architects, such as John W. Root, Raymond Hood, Ralph Walker, and Ely Kahn, and such talented foreign architects as Eliel Saarinen, Joseph Urban, and Eugene Schoen. It was, therefore, an establishment show. Every object in the exhibition—furniture, accessories, and furnishings—was designed by the architects or produced by a few designers serving essentially as artisans. Peter Muller-Munk, for example, was permitted to design and execute a silver tea and coffee service for Ely Kahn's outdoor dining room, and Ralph T. Walker commissioned Egmont Arens to produce a fountain of light and Maurice Heaton to provide decorative glass for his sales-room. Richard Bach was evidently determined that the leadership of the new profession be vested in architecture.

Despite the great popularity of the Metropolitan show, there were some critics who felt that the public had come more out of curiosity than out of conviction and that the modern movement had yet to prove itself. Royal Cortissoz, in a *Scribner's* article entitled "A Contemporary Movement in American Design," questioned its dependence upon foreign talent and its "strange conformity" with foreign motivation. He was convinced that the Americans were entering an age of speed and that they were, therefore, in "more a transitional than a decisive period." Cortissoz recognized that "there must be acknowledgment of the fact that the machine, as a tool of the designer, has replaced the craftsman in contemporary production, and has, therefore, tremendously influenced modern design," but felt that there was a certain shallowness in the tendency to "go modern" as one might "take up golf." (124, 595) Cortissoz found the modern interest in straight lines appropriately hard and crisp but not human. He lamented the bleak rigidity of some modern design and the "Hollywood de Luxe" style of others. In a curious way the modern style seems to have been regarded by many as being more suited to the threatrical and motion picture worlds. It is an interesting anachronism that the important theaters and the grand movie palaces of the period were extravagant palaces in traditional styles while the plays and

motion pictures themselves most often dealt with modern themes in modernist settings.

By the end of the 1920s manufactured products had followed electricity into the American home. The radio had taken over the living room, and the refrigerator dominated the kitchen. Major manufacturers had accepted the inevitable fact that the appearance of a product—its styling in a more fashionable form—was indispensable to its success in the marketplace. In 1928 the advertising and printing trade publication *Printer's Ink* published a timetable of those professions that its editors believed to be the most important to the progress of industry in the United States. For 1900 to 1910 they gave the greatest credit to salesmen, and for 1910 to 1920 to advertisers. Then they proposed that the honor for 1920 to 1930 should be given to artists in industry.

The American Management Association climbed aboard the design bandwagon in 1929 by conducting a series of seminars for industrialists and businessmen on whether "intelligent and profitable use [can] be made of the modernistic style of design." (211, 4) One seminar considered "How the Retailer Merchandises Present-day Fashion, Style and Art." In another seminar, H. S. Nock, Paul Bonner, and John Alcott (then head of the design department of the Massachusetts School of Art and Design and design consultant to the Associated Industries of Massachusetts) reviewed "How the Manufacturer Copes with Fashion, Style and Art Problems." And in the third seminar, Ralph Abercrombie lectured on "The Renaissance of Art in American Business."

On the very day of the stock market crash, October 29, 1929, the American Management Association was holding its annual convention in Detroit with E. Grosvenor Plowman, advisor on merchandising problems to the Associated Industries of Massachusetts, as the keynote speaker. "There was a time," he said in his talk, "when our best things were hand-made, our poorest made in mass production. Cheap, nasty, poor taste things were turned out by the machine. The reverse seems to be beginning to be true

The Metropolitan Exhibition of Contemporary American Design, called "The Architect and the Industrial Arts," included this backyard garden by Ely Jacques Kahn. The furnishings were by Kahn with the artisan Walter Kantak, and the silver tea and coffee service was by Peter Muller-Munk. American Magazine of Art, April 1929.

Eliel Saarinen's dining room at the Metropolitan show included furnishings and accessories designed by the Danish master himself, though executed by others, as well as fabrics designed and made by his children. The Architect and the Industrial Arts (Metropolitan Museum of Arts catalog), 1929.

today." (211, 16) The question-and-answer period that followed was particularly revealing of the current attitudes toward design. To the question of how much beauty should be put into products, Plowman responded that one must "sell beauty to the public." The question of the conflict between traditionalists and modernists drew the response from him that the so-called modernistic style was the "offspring of the jazz age," that "the modern style will remain as long as the machine in its present form is the dominant characteristic of world production, [changing] its outward form from year to year, just as women's dress styles change," and that manufacturers would have to learn to balance faddism and permanence. (211, 4) The discussion that followed brought out the significant opinion that "almost all industries have an overcapacity from a production standpoint," and that therefore "there are more goods than there are consumers" and all manufacturers in any given field were attempting "to sell much the same articles to the same consumer." It was claimed that this situation was responsible for the emphasis on appearance for the sake of appearance. To the classical question of the difference between good and bad taste, Plowman responded that good taste was "outward evidence of inner culture and refinement," and made a particular point of the fact that, despite European criticism of American products, they were moving toward "good design to fit in American life." (211, 6) He concluded the session with the observation that the day of period styling was over and that the manufacturers would have to pay attention to the modern ensemble styling, in which every element in a group was turned to the whole.

In general, the attendees at the AMA meeting agreed that art values had become essential to sales success. It was pointed out that, although the works of a simple radio cost only $50.00, "in an artistic cabinet it brings $150.00 currently, or $750.00 from the well-to-do buyer." The only reference to education at the meeting was that schools were developing "a generation who is thinking in terms of appearance before cost and are not interested in durability at the price of ugliness. . . . It is art, not money, that 'makes the

mare go.'" (211, 31) Other participants also attested to the value of product aesthetics. Stanley Needhouse of the L. C. Smith Typewriter Company noted that his company had introduced simple colors in its products in 1926 after years of shiny black. By 1929, he noted, only 2 percent of its typewriters were still in black and the green "Secretarial" model had become common in business offices. E. B. French reported that Egmont Arens's 1928 redesign of his company's Kitchen-Aid mixer, brought out originally in 1921, had cut its weight in half, reduced its price and improved its appearance; as a result, sales had increased 100 percent. One participant emphasized "the need of combining mass production with style appeal in order to sustain the structure of American business." Another person said, with enthusiasm: "We are going through the golden age of modern industry. We are in the empire period of mass production. . . . Art in industry is the natural and logical climax of a century of mechanical invention and industrial progress." (211, 16) A more restrained opinion, from a man whose company manufactured noncompetitive products, favored industrial design not to increase sales but out of recognition "that the modern trend is toward a more pleasing appearance of utilitarian things." (211, 28)

It was at this AMA conference that the specter of deliberate product obsolescence appeared publicly for the first time, as a calculated reaction to the public's affection for up-to-date design. It was debated whether "the encouragement of progressive obsolescence was a logical function of the stylist or the designer." Lewis Mumford had already warned elsewhere in 1929 that the furniture industry had a goal of replacing furniture every 6 years by changing styles deliberately or by building products with limited durability. Plowman's position was that the rapid pace of inventions in the United States had been originally responsible for inducing obsolescence of products before they were very old. However, now, "the burden of causing obsolescence of articles before they were worn out [had] to be transferred in part to the designer because . . . the mechanical development of certain things]had] reached an approximate upper level of development." (211, 14)

As the United States plunged into the Depression, it became evident that the public could be stimulated to spend its hoarded and often limited money on a manufactured product only if it took on an entirely new appearance—one that was more than a simple shift in style and that promised better times ahead for those who had faith in America and American industry. This state of mind gave product design a sense of responsibility that helped to establish industrial design. Despite the once prevalent attitude that it was somehow disrespectful if not indecent to use aesthetic values for product promotion, it now became apparent that vernacular manufactured products had aesthetic dimensions that in their own way were as valid as were those once reserved for exclusive aristocratic products.

The House of Tomorrow

*There is nothing in the nature of machinery
which requires that these things be aesthetically
inferior. Nor is the American love of the machine
demonstrably a barbaric trait.*

Frederick P. Keppel, 1933 (53, 28)

Early in 1929, President Herbert Hoover, perhaps
with some premonition that the euphoria of the
Americans was soon to be shattered, had appointed a Research Committee on Social Trends
"to throw light on the emerging problems which
now confront or which may be expected later
to confront the people of the United States." (53)
When one of the monographs of the committee
was published in 1933, after the first deep shock
of the Depression and the replacement of Hoover
by Franklin D. Roosevelt, it made particular reference to the importance of art and design in
American life and found cultural as well as economic significance in the observation that streamlined products seemed to most people to be
better-looking and more appropriate to modern
times. The cultural lag that had been given as
an excuse for America's inferiority in the area
of aesthetics seemed to have been dissipated,
at least in the area of utilitarian products, by
a shift in the public temper. People now looked
forward hopefully to the new forms of airplanes,
automobiles, and other miracles of modern
technology.

The failure of the stock market had brought down
most of the grand dreams and fortunes of the
1920s, but not those of the designers and stylists
of manufactured products. Industrial design came
into its own in the decade between the Depression and World War II. While other countries were
abrogating their pledges to the Armistice of 1918,
the United States was recovering its economic
momentum and discovering a cultural force it had
not recognized before. For a few years at least,
before the United States was drawn into the war,
Americans were free to develop their talents in
the unique arts of the man-made environment.
Skyscrapers reached for the sky, airplanes raced
the sun, automobiles put everyone on wheels,
and homes promised to become a mechanical
and electrified paradise as both old and new
products were refined and preened in competition

for the consumer's attention. In the process, industrial designers became the glamor boys of the
moment, accountable not only for mass-produced
aesthetics but also, before the decade was out,
for the machine's conscience. Ugly and clumsy
products were to be no longer tolerated just
because they worked. It was expected that they
would be considerate of the public's psychological as well as physical well-being.

The American Union of Decorative Artists and
Craftsmen (AUDAC) had been formed in 1927
by a group of decorative and industrial artists
attempting to protect themselves from the piracy
of their concepts by manufacturers who had long
been accustomed to borrowing traditional designs with little fear of reprisal. Now that their
livelihood was threatened by such practices, they
proposed that such borrowing "degrades the
producers [and] corrupts the taste of the public."
The stated purpose of the AUDAC was "to cooperate with manufacturers and the public in the
placing of American arts and crafts on a basis
of honesty, dignity and merit." (35, 187) Moreover,
its members proposed that without such concentrated and patriotic effort, no style truly American could be expected to mature into fruition.

After the Metropolitan Museum's architect-
dominated 1929 exhibition, the AUDAC hastily
organized and staged its own exhibition of five
modern ensemble interiors by members in early
1930 at the Grand Central Galleries in New York.
The objective was to demonstrate the American
potential for original design in the decorative and
useful arts and to prove that the general public
could be served just as well by the well-designed
products that were already on the open market
as was the upper class by the custom-designed
and specially produced products in the Metropolitan's exhibition. Frank Lloyd Wright, Lewis
Mumford, Norman Bel Geddes, and other prominent design personalities of the era contributed
statements to the *Annual of American Design,*
which was published in 1931 by AUDAC to promote the aesthetic and creative competence of
its members and to demonstrate that human
comfort and utility were proper goals for Ameri-

can business. They were convinced that the vernacular products of industry were destined to acquire cultural as well as social significance. Bel Geddes, in his article, praised the successful cooperation between art and industry. On a different tack, Mumford warned that manufactured products should be based on the principle of conspicuous economy rather than either Veblen's conspicuous consumption or the concept of conspicuous waste that lay just over the horizon of plenty. Their action was recognized by others as a demonstration of the potential for a vital renaissance of the useful arts and as evidence of their conviction that modern life deserved a more appropriate and useful setting.

Although the department stores and other merchandising outlets had been second only to the advertising agencies in introducing design and Art Moderne to the public and in guiding and molding public taste, when the Depression moved to its deepest point their interest in modern design waned. American consumers were no longer fascinated with its stylistic appeal and refused to buy it. In fact, in some circles the French Art Moderne style, with its appeal to high fashion and machine-age mannerisms, was blamed for the economic catastrophe. The department stores reverted in large measure to their original strategy of catering to public taste rather than trying to guide it. The stores that had become cultural showcases, offering exhibits, lectures, and concerts to the public and blazing a trail as missionaries of modern aesthetics, now abandoned that responsibility to their buyers, whose preoccupation with certain profit blinded their cultural sensitivity. Many of the department stores that were once the patrons of good design have since deteriorated to the raucous merchantry of discount houses.

However, the modern style, at least in the industrial and decorative arts, had become dependent upon new materials and new manufacturing techniques. Modern wooden furniture depended upon veneers of exotic woods, such as rosewood, zebra wood, or ebony, applied to rectilinear solids in a way that concealed the substructure of mixed and more humble materials. Some claimed that this was the age of metals. "No

metal, it seems to me," testified Emily Post in an advertisement in *House and Garden,* "is quite so complete an answer to the housewife's prayer as chromium—appealing not only to the eye, but to practical requirements." (144, 2) Aluminum, which had been used for machinery and as a skin laid over the wooden frameworks of custom-made automobiles, now found a more dramatic application as the stressed skin of modern aircraft and was, at last, being used effectively for kitchen and gift wares and other modern furnishings. And Monel metal, a copper-nickel alloy that was a precursor to stainless steel, was popularized for domestic and commercial sinks and cabinets. Black "Vitrolite" glass became the standard material for storefronts. And mirrors, more often than not round, became the icons of the modern interior. New composition materials, like Masonite, cork, linoleum, and rubber flooring, prepared the public for the plastic age.

The Depression pushed the infant plastic industry into a cycle of growth that continues still. Celluloid (1870) and Bakelite (1907) had been around for decades, but the flammability and instability of the first and the drab colors and manufacturing difficulties of the second had limited their application. However, in 1932, when the General Mills Company conceived the idea of stimulating interest in its new breakfast cereal by offering a plastic "Skippy Bowl" as a premium with each purchase of two boxes of Wheaties, the plastic boom was launched. Skippy was the movie name of an endearing child actor, Jackie Cooper, who had just appeared in a film with Wallace Beery called *The Champ* and had been taken to heart by the public as a victim of the Depression. The premium bowl, made out of a urea formaldehyde thermosetting plastic produced primarily by the American Cyanamid Company, was useful, hygienic, colorful, and virtually unbreakable. Over 5 million were distributed. Other plastic giveaways followed, including 10 million biscuit and cookie cutters that were distributed to publicize the prepackaged dry dough mix Bisquick. In time, the services of many designers would be required to transform products that had been made of tra-

Chrome-plated coffee and tea service designed by Ilonka Karasz in the Art Moderne style, with geometric forms and no decoration. Reference 35.

Lurelle Guild's work with
the Wear-Ever Aluminum
Company resulted in a
long line of modern cook-
ing utensils. In particular,
his cast aluminum kettle
has become a typeform.
Lurelle Guild.

Some of Russel Wright's
1932 line of lathe-spun
aluminum ware. The prod-
ucts were designed to be
taken, as he said, from
stove to table. Russel
Wright.

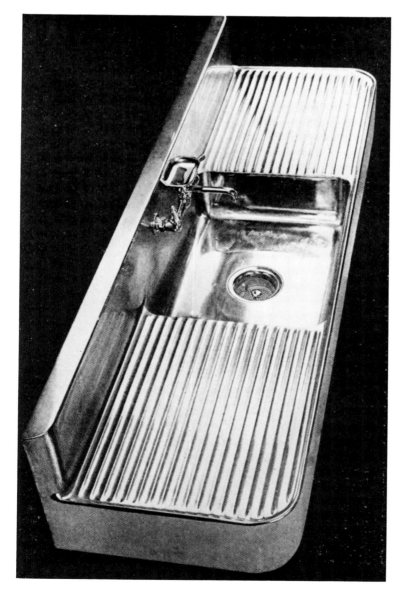

Gustav Jensen's hand-some Monel-metal sink, designed in 1931 for the International Nickel Company, reflects his classical sense of proportion and form. Pencil Points, *March 1937.*

Gustav Jensen was one of ten industrial designers who were each paid $1,000 by the Bell Telephone Company to come up with the ideal dial telephone. The model he submitted was in sharp geometric style of the era. However, the account went to Henry Dreyfuss, whose solution paid less attention to style than to function. Pencil Points, *March 1937.*

One of Jensen's most dramatic designs was this spun aluminum radio case. Its innovative geometry broke with the conventional cabinet form. Though it was never manufactured, its modernism influenced other designers and manufacturers. Pencil Points, *March 1937.*

This apartment by Joseph Urban shows the rather pristine classicism of the Art Moderne style at its best. Decorative Arts, 1933.

A 1933 illustration by Gilbert Rhode for an advertisement promoting Sealex linoleum. House and Garden, *September 1933.*

*George Sakier's faucet
set for American Standard
has been called the best
of the functional modern,
despite Sakier's personal
conviction that function
follows form.* Arts and
Decoration, *November
1933.*

ditional materials for plastic production. "I have always felt," wrote Gage Davis, president of the Spartan Corporation, "that this mass 'sampling' of a relatively unknown material helped to introduce the entire emerging plastics field to the American consumer market. . . ." (207)

The popular acceptance of new materials as a promise of things to come helped to offset the negative reaction to Art Moderne and the decline of faith in its obsession with geometry. The adaptation of new materials from their basic sheet, bar, rod, and tube forms tended to restrict, especially in custom-made products, the application of decorative detailing. The simplicity of form that this encouraged was perceived as not only more beautiful but also more useful. Some designers, however, protested the implication that plainer and more geometric forms were more functional. George Sakier, director of the Bureau of Design of the American Radiator and Standard Sanitary Corporation, was convinced that designers were being held back by the cult of functionalism. And Kem Weber found a basic conflict between formalism and functionalism: "Design . . . is the effort to express the basic form, as such, aesthetically truthful to material and purpose." (35, 13) Paul Frankl, also a formalist, accepted the idea that modernism was "the style of reason." "Its appeal is an appeal to the intelligence," said Frankl." "Its emphasis upon simple forms, its return to mathematical axiom and the fundamentals of form, confer upon it a classical rather than a romantic beauty." Nevertheless, Frankl was convinced that it was wrong to assume that everything that is useful is beautiful: "Usefulness and beauty in themselves are mutually unrelated terms. Beauty is concerned with form only; usefulness only with serving a purpose. The fact that the form of an object is derived from its function has been adopted as one of the fundamental principles upon which the contemporary work is based. The idea is not new, nor has it anything to do with the art creating of this or any other period. . . . Creative work is an impulse which begins where the reasoning of the mind leaves off. . . . The trend of the day is toward mass production. Comfort, livability, soundness of construction, moderate prices—these are the desiderata of the newest Americana." (35, 13–83) To Walter Dorwin Teague simplicity in products also implied more than formalism; it was an attempt to make manufactured products more comprehensible to the consumer. Russel Wright felt much more strongly that the functionalists had, by their interest in mechanical form, been led into ignoring the human form.

In 1930 the Metropolitan Museum staged its third International Exhibition of Contemporary Industrial Art, this time with an emphasis on metalwork and textiles. It was to be the last in a series organized by the American Federation of Arts (F. A. Whiting, president) under a grant from the General Education Board, for the purpose of "demonstrating design in current production." This was, again, a Europe-dominated show, with only a small array of works by Americans like Paul Lobel, Russel Wright, Donald Deskey, and Walter von Nessen and émigrés like Eliel Saarinen and Peter Muller-Munk. There were entries from England, Holland, Sweden, Switzerland, and Czechoslovakia. The Germans sent, for the first time apparently, metalwork by such former teachers and students of the now-closed Bauhaus as Walter Gropius, Marianne Brandt, Wilhelm Wagenfeldt, and Hin Bredendieck. In the exhibition catalog Charles Richards, former director of industrial arts of the General Education Board and now executive vice-president of the New York Museum of Science and Industry, acknowledged the modern movement and called for a machine aesthetic independent of that represented by the modern movement in art. He concluded that the new spirit of design was entirely suited to the machine, having emerged as the result of social and economic change. It was an aesthetic that was particularly fitting to those products that had "come into being in the industrial age with no traditions of craftsmanship behind them," such as the automobile. "We design," said Richards, "wholly with reference to the machine, and the appearance of these things represents what the machine can most readily, naturally, and effectively produce." (194, 60) Catherine Louise Avery, curator at the Metropolitan, commented in the Museum's bulletin that modern European craftsmen were stressing "functionalist design," defined as design determined by the process of manufacture and end function. "The results," Avery wrote, "are often severe and uncompromising, but in their very insistence they gain their point. People numbed by seeing nothing but conventional patterns, used to superfluous and stupid ornamentation, can perhaps be roused from this apathy only by the strong medicine of German and Scandinavian expression." (153, 263) Richards wrote: "It is the Germans who carried this idea farthest. With characteristic zeal, they are concentrating upon the effort to produce 'type forms' in which both the limitations and capabilities of the machine are recognized and which can be produced with the greatest speed and economy." (195, 609)

A year later, the Metropolitan staged the twelfth American industrial art exhibition in the series that had begun in 1917. The show had now broadened its scope to include what some now called the art industries (ceramics, tablewares, textiles, lighting fixtures, and furniture) as well as the "artless" industries (major and minor manufactured appliances). Its primary purpose was stated as intending to trace the effect of the 11th exhibition of 1929, whose interiors and furnishings had been custom-designed by architects. Of course, the practice of designing for manufacture was already well established outside the sphere of influence of museums. Nevertheless, acting in their self-anointed role of preachers of taste, museums had demonstrated their power to canonize an aesthetic movement or a style. Over the years since the exhibitions began, the sacred relics of tradition had been converted to Art Moderne, lost to the purgatory of modernistic, and then elevated to modern, and now R. F. Bach proposed that the new religion of design should be baptized as *contemporary.* Hence the exhibition was given the new title of Contemporary American Industrial Art. Henry Dreyfuss concurred: ". . . good modern design does not have to be 'modernistic'. ('Modernistic' is rightly used to describe the odd-looking objects thrown together to look extreme and outlandish, while the word modern or 'contemporary' denotes the spirit of today interpreted in good taste.)" (126, 192)

Still, museums were as reluctant then as they are today to recognize mass-manufactured products on their own terms against their own environments. It was believed that to be evaluated properly such products had to be viewed "without the insistence and limitation of salesroom and counter, without the interference of captious customers, without reliance upon advertising and selling talk." (154, 226)

The secular world, however, saw the exhibition in a different light. *Business Week* magazine found it a sharp contrast to what it called the "gaudy" 1929 exhibit, and welcomed the attention to products "designed by Americans, made by repetitive processes or machines" away from the "exhuberance and easy money" of the past. (119, 22) And a *New York Times* editorial recognized that the exhibition held "encouraging promise to those who, while cherishing the past, watch for new combinations of usefulness and beauty." (190)

Whereas the industrial arts had been promoted in the 1920s as a means of achieving status by way of manufactured luxuries, industrial design was more concerned in the 1930s with making common necessities attractive to the general public. The Great Depression of 1929–1935 may have provided the catalyst Americans needed in order to recognize that there was beauty and à natural elegance to be found in vernacular products. It became evident that domestic appliances and business and industrial machines could be sold if they were endowed with good proportions, fresh colors, and modern detailing. Then in 1932, during the brief period when prices were stabilized under the National Recovery Act, manufacturers recognized that, quality and price being equal, a product's appearance became paramount in attracting the buying consumer. New models of old products as well as completely new products provided the stimulus that helped the economy turn upward again. In this respect, Norman Bel Geddes may have been uniquely instrumental in stimulating public aspirations for a better future. His book *Horizons* proposed a future of hope and progress with design as the great liberator, and his bold prophecies in a 1932

WE DO OUR PART

The symbol displayed
by companies cooperating
with the short-lived Na-
tional Recovery Adminis-
tration. Author's collection.

issue of *Ladies' Home Journal* reflected the
optimism and vision of the twentieth-century
entrepreneurs. He foresaw a time when, among
other things, a new fuel of vastly improved power
but infinitesimal bulk would replace gasoline, synthetic materials and curtain walls would dominate
buildings, photoelectric cells would open doors,
and the garage would move in to face the street
and become part of the house. He was convinced
that artists would be "thinking in terms of the industrial problems of their age," and that utilitarian
objects would be "as beautiful as what we call
today 'works of art.'" (140, 3)

In 1933, with the country still in economic trouble,
the Century of Progress Exposition opened in
Chicago. It was a neon, krypton, and xenon celebration, with lineal decorative lighting provided
when tubes containing these rare gases were
energized. For the first time industrial designers
were provided with the unique opportunity to
meet the challenges of design for the larger
man-made environment. Norman Bel Geddes,
who had served since 1928 as a consultant on
lighting to the Architectural Commission for the
Fair, had developed, with his associates, a series
of innovative theater and restaurant concepts—
including the first proposal for a revolving restaurant perched on a high tower, a feature that
has become almost mandatory for high-rise
hotels and television towers. Although Bel Geddes's imaginative suggestions ended up as
casualties of the Depression (perhaps because
of the architects' reluctance to accept suggestions from a nonarchitect), they broke dramatic
new ground by suggesting that theatrical settings
and the real world could be joined in an exciting
human experience. For many visitors the Century
of Progress Exposition offered a dream world
where one could escape the austerity of a trou-
bled time by living for a few hours in the future.

Other designers whose recommendations contributed to the fair included Joseph Urban, who
provided a palette of harmonious colors for the
buildings; John Vassos, who designed the unique
and now classic Perey turnstiles; and Wolfgang
Hoffman, Jean Reinecke, Gilbert Rohde, Eugene

The symbol of the Century of Progress Exhibition combined a spinning Earth with the heavy graphics that were considered modern at the time. Official Guide Book to the Fair, *1933*.

The Travel and Transport Building at the Century of Progress Exhibition, called "a sensational piece of modernistic fair architecture," made no effort to hide the circle of cables and steel towers that held up its roof. Its structure echoed Buckminster Fuller's Dymaxion House patent of 1925. Applied Photography, *October 1934*.

A number of buildings at
the Chicago fair, including
this General Motors structure, were in the sky-
scraper style. However,
they were conceived as
solid sculpture rather than
as functional shells. Ap-
plied Photography, Oc-
tober 1934.

Norman Bel Geddes's dynamic proposal for an "Aerial Restaurant" for the Chicago fair. The entire superstructure, with its three counterbalanced dining areas, was to revolve above the kitchens set in the base. Hoblitzelle Theatre Arts Library, University of Texas, Austin.

This "Island Dance Restaurant" proposed for the Chicago fair was to accommodate 1,500 people on four islands in the center of a lagoon. Hoblitzelle Theatre Arts Library, University of Texas, Austin.

Another Bel Geddes proposal for the Chicago fair: an "Aquarium Restaurant" to be located in the wall of a dam. Hoblitzelle Theatre Arts Library, University of Texas, Austin.

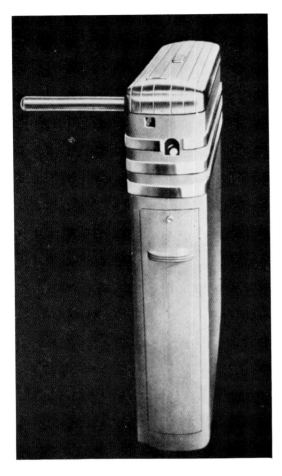

John Vassos's turnstile for the Perey company, first used at the Chicago fair, became a standard in the industry and is still in use at some airports and other public places. John Vassos.

Schoen, and Walter Dorwin Teague, who served as consultants to several of the exhibitors. Teague's Rotunda for the Ford Motor Company emerged as one of the hits of the fair and was subsequently rebuilt at the company's headquarters in Detroit—this time with a Fuller geodesic dome.

The faith of the Americans in the future was well served in this difficult time by the Chicago fair's "House of Tomorrow," a circular structure (not unlike Fuller's Dymaxion house) designed by George Fred Keck of Chicago with a steel frame and glass walls around a central column for utilities. The House of Tomorrow had a sun deck, a garage for an automobile, and a hangar for an airplane. Its innovative rooms were displayed *"en ensemble"* with such a high degree of believability that the public accepted them as the logical and, therefore, inevitable fruit of technology.

Thus American designers began to demonstrate their ability to reach beyond the short-term commitments of manufactures to a position of design leadership that would serve to quicken the pace by which science and technology could be transformed for human service. The American public found hope in the Design Decade (as the 1930s had been aptly named by *Architectural Forum*) and began to look upon its modern manufactured forms and newly styled appliances for bathrooms, kitchens, and laundries as indispensible to living in the modern world. One observer of the time warned that the imagination of men like Norman Bel Geddes would "cost industry a billion dollars" because such daring ideas threatened to obsolete existing products and factories by stimulating public aspiration for better things. There was excitement in the air as the American Dream gathered momentum again.

The Chicago fair's "House of Tomorrow." Chicago Historical Society.

Skyscrapers and Streamliners

. . . the new office of industrial designer can claim no superiority over the well-trained architect.

Architectural Forum, December 1934 (94, 409)

Once again, in 1934, the Metropolitan Museum held an exhibition of industrial art in modern home furnishings. However, despite the pressure on the museum to avoid "the self-consciously clever design of five years ago, supported by an economic scheme only a little more false than its accompanying social concept" (as it was put by *Architectural Forum*), the museum's insistence that all of the objects in the exhibit be shown for the first time produced things that were again out of context with the economic conditions. The museum was still preoccupied with the notion that all design should stem from architecture, and thus most of the exhibits were commissioned from architects. In recognition of the growing importance of industrial design, however, a fair percentage of those invited to participate were professional designers, including Walter Dorwin Teague, Raymond Loewy, Donald Deskey, Gilbert Rohde, Gustav Jensen, and Russel Wright.

It was inevitable, perhaps, because of the sympathetic attention of museums and the vested interest of the architectural press, that the popularity and potential of industrial design should attract the attention of young architects who were finding few architectural commissions and who had an inclination toward a broader application of their design talent and training. Montgomery Ferar, Dave Chapman, Ray Sandin, and Brooks Stevens were among the early architects who, followed by George Nelson, Eliot Noyes, Charles Eames, Walter B. Ford, and others, developed successful careers in industrial design. The range of their feelings about their career shift is reflected in Sandin's "I got into this profession by accident and I am most happy that I did" (218) and Chapman's "Frankly, I still thought of myself as an unemployed architect, pacing out the depression in an attractive, lucrative but, nonetheless, substitute career." (217)

Over the years, design and architecture have enjoyed a warm but wary relationship. Architecture has largely ignored its upstart friend, yet has turned to it either in times of economic stress or in the search for that instant fame that comes with having one's name attached to a unique chair. Design, on the other hand, has often hungered for the status of architecture (Walter Dorwin Teague doggedly pursued an architectural degree until he acquired it in 1938 at age 55) or looked for and caught the flame of the latest formalistic fashion from its senior associate.

Further evidence that architecture was reaching out for its share of the influence and affluence of industrial design is that, coincidental with the Metropolitan show in 1934, the young Museum of Modern Art, with Alfred H. Barr, Jr., as director and Philip C. Johnson as director of the department of architecture, installed an exhibition on "Machine Art." Whereas the Metropolitan was catering to the personal environment and was preoccupied with formalism as personal expression in the current fashion of modern art, the Museum of Modern Art elected to meet human needs with mechanical means and found its formalism in the geometry of solid shapes—forms so mathematically pure that the object "loses all character and distinguishing marks of purpose," as Geoffrey Holmes observed. It is evident that the theme of the MOMA exhibition was stretched to include the categories of geometrical shapes in household equipment and furnishings carefully extracted from the work of industrial designers who were also represented in the Metropolitan show with more personal and humane forms.

Alfred Barr was quite correct in pointing out that a product has a mechanical function (how it works) as well as a utilitarian function (what it does). Years earlier, however, as a teacher of art, he had invented for his students the wise game of having them search the outside world for well-designed objects selling for less than $1 in order to stimulate their sense of value. It is difficult to understand why he did not include a humanistic function (the why). Phillip Johnson properly objected to what he called "French-age aesthetics" and to the irrelevant styling and irresponsible streamlining that were gaining popularity. However, there still remained his presumption that solid geometry can be equated with utility.

Walter Dorwin Teague's water glasses for Corning, conceived as pure cylinders, were included in the Museum of Modern Art's Machine Art exhibition in 1934. Museum of Modern Art.

This set of stainless steel
mixing bowls manufac-
tured by the Revere com-
pany was included in the
Machine Art exhibition.
Revere Copper and Brass
Co.

The third attention-getting exhibition of 1934 represented the final side of the philosophical triangle of industrial design. If the other two sides were the formalisms of personal and impersonal aesthetics, the third was the formalism of public aesthetics. The exhibition, entitled "Art in Industry," was held at New York's Rockefeller Center (whose Radio City Music Hall, incidentally, now a national monument, had been designed by Donald Deskey). It was organized by industrial designers under the sponsorship of the National Alliance of Art and Industry to illustrate, according to the prospectus, "what designers are doing in the way of conscious creation of forms to help the engineer sell his mechanical devices." (93, 331) Once again the work of industrial designers, like Walter Dorwin Teague, Russel Wright, Gilbert Rohde, and Gustav Jensen, was included in the exposition together with about a hundred of their less famous colleagues showing more than a thousand designed products. The exposition provided a glittering showcase for many of the giants of the young profession. However, their very success led to a complicated situation that was eventually to draw industrial designers into a professional society.

The following year, 1935, when the National Alliance of Art and Industry decided to hold another exhibition, its officers decided to turn responsibility over to the manufacturer members of the Alliance rather than the designers. Many of the most prominent industrial designers objected to their having no part to play in the quality and contents of the exposition. Thus, when President Roosevelt tapped a golden telegraph key in the White House to open the showing in New York and Fiorello LaGuardia gave his opening speech praising the design of utilitarian products, examples of the work of these outstanding designers and their important clients were missing from the exhibits. Even the showing of a model of Frank Lloyd Wright's "Broadacre City" could not cover the fact that without their presence and aesthetic guidance the show was a carnival of bad taste and irrelevant exhibits.

The designers who had boycotted the show, feeling their strength, issued a manifesto stating that the Alliance "neither stimulates better de-

sign, represents the artist, nor improves the relationship between the designer and industry" and that "it does not promote the best standards of American design." (191) As a result the leading designers, together with other interested designers, got together in a protest meeting. They compared their grievances against the Alliance and promised themselves to form their own organization soon.

With these exhibitions and others it became evident that the concept and promise of industrial design had caught the public imagination and that designers would be hailed as the heroes of the economy. Some magazines even suggested that designers may have been instrumental in pulling the nation out of the Depression. Designers became popular subjects in the Sunday newspaper supplements and the focus of innumerable stories in the business and general magazines. Editors of periodicals aimed at the higher classes discovered the subject of domesticated aesthetics and devoted themselves to articles on a higher and better life surrounded by the elegant vernacular of modern industry. In an article entitled "The Eyes Have It," *Business Week* declared that " 'stylizing' has become an effective weapon in meeting new competition." (118) *Forbes* magazine, under the title "Best Dressed Products Sell Best," suggested that "progress, profit and patriotism do mix." (128) And the pamphlet "Dollar and Cents Value of Beauty," published by the Industrial Institute of the Art Center in New York, accepted industrial design as "not a luxury, but an economy; not a fine art, but a practical business" and extolled the profit value of bringing out new models to whet buying interest. (215) On this last point, over which controversy raged for half a century, it must be remembered that planned obsolescence seemed to make sense at a time when the economy needed a jolt to get it moving again. The idea was given form by Ernest Elmo Calkins in 1930— in part, perhaps, as a tactic that would be beneficial to his own business of advertising: "The styling of goods is an effort to introduce color, design and smartness in the goods that for years now have been accepted in their stodgy, commonplace dress. The purpose is to make the customer discontented with his old type of fountain pen, kitchen utensil, bathroom or motor car, because it is old-fashioned, out-of-date. The technical term for this idea is obsoletism. We no longer wait for things to wear out. We displace them with others that are not more effective but more attractive." (120, 153)

The major overview of the design profession in the early 1930s seems to have been *Fortune* magazine's "Both Fish and Fowl" (129), which scored the failure of industry to produce new designs as the chief cause of the persistence of the Depression and attributed the fact that the nation's economy was on the rise again to industrial designers. The article attributed the 900-percent increase in sales of the Toledo Company's "Public Health" scale to its redesign by the firm of Harold Van Doren and John Rideout. Donald Dohner's design was credited for a 700-percent increase in sales of a Westinghouse range, as was Raymond Loewy's design of a radio manufactured by the Colonial Company. There are many other examples of market recovery for products after the doctors of manufactured aesthetics had applied their healing and renewing prescriptions.

Norman Bel Geddes's most spectacular and perhaps only real success in industrial design was his transformation of the Standard Gas Equipment Company of New Jersey's series of over 100 models of stoves into 12 standardized components that could be recombined to produce 16 different models. Bel Geddes developed a structural system that did away with the old method of bolting cast and pressed plates together in favor of a rigid skeletal frame upon which the various components and plates could be hung in a system not unlike the curtain wall structures that were becoming standard in architecture. His redesign saved the client thousands of manufacturing dollars and established a type-form for large appliances that is still standard in the field. Bel Geddes's public status at the time was such that the client was able to capitalize on it by attaching his monogram to it as an added buying attraction—just as prestigious and fashionable products are emblazoned today with contemporary design stars' marks.

Henry Dreyfuss opened his own design office in 1929, when he was 25 years old, and although he continued to design stage settings for a while he gradually turned to industrial design. The Westclox Big Ben and Baby Ben alarm clocks were early Dreyfuss successes; in styling and packaging they set a new direction for the clock industry. In 1932, Dreyfuss was contracted by the Sears, Roebuck Company, in its first venture into industrial design, to redesign its wringer-type washing machine. Dreyfuss enclosed the tub and motor of the machine with a metal skirt painted with a textured green enamel and held in place with the then-popular chrome bands, which also served to mask the connecting bolts. In addition, he convinced the manufacturer to move all of the machine's controls to the top in order to make the washer simpler and safer to operate, and named the machine the "Toperater" in order to emphasize that particular advantage. Dreyfuss added his escutcheon to the product for status. The machine was a runaway marketing success for Sears.

A year later, when Henry Dreyfuss was under contract to the General Electric Company to redesign its "monitor-top" refrigerator, he felt obliged to turn down a second contract from Sears to redesign its line of refrigerators and recommended his colleague Raymond Loewy. Loewy took on the contract for $2,500, although he claims that it cost him three times as much to finish the job. Loewy's design was the first Coldspot in a line of successful refrigerators whose sales grew from 15,000 to 275,000 units within five years. The acceptance of these first two major appliances established Sears as an important competitor in the major appliance field and convinced management to employ Jack Morgan to organize the company's first industrial-design department. Furthermore, the Coldspot launched Raymond Loewy on the road to success and eventual international stardom. His first employee was Robert Jordan Harper, who joined him in 1932 at the grand Depression-era salary of $10 a week. Harper worked with Loewy until 1935, when he joined Walter Dorwin Teague, who at that time had six employees.

Norman Bel Geddes designed this stove in 1932 for the Standard Gas Equipment Company. The twelve standardized elements could be recombined into sixteen models. The structural concept followed the skyscraper method: the white enameled panels were attached to a steel skeleton. U.S. Patent 90,108, June 13, 1933.

The Bel Geddes stove attempted to set a standard for the dimensions of kitchen appliances, cabinets, and counters. Arts and Decoration, November 1933.

The alarm clocks that Henry Dreyfuss designed for the Westclox company demonstrate his sensitivity to public taste. The clock forms were a break from the traditional form with the bell on top, yet were reminiscent of the mantle clock form. They were weighted to compare well with the older clocks. Also, the packaging was in the moderne pattern of parallel lines. Design, *January 1935.*

The "Toperator" washing machine was designed by Henry Dreyfuss for Sears in 1933, at the height of the battle for the washing machine market. The body, done in mottled green enamel, was trimmed with chrome bands and carried a Dreyfuss nameplate. Dreyfuss Archives.

This Coldspot refrigerator, designed for Sears by Raymond Loewy in 1934, was developed with particular attention to the consumer's needs and moved Sears from below tenth place to one of the top three companies in the industry. Raymond Loewy.

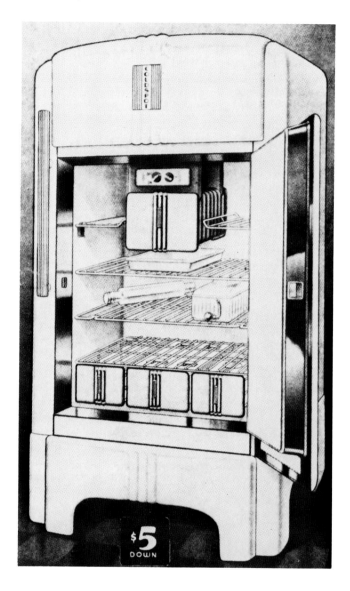

One of the most handsome products of the 1930s, and the first automobile to reach production after styling by an industrial designer, was Walter Dorwin Teague's Marmon 16. To some extent the form of the Marmon may be credited to Teague's son, Walter Dorwin, Jr., who, while a first-year student at Yale, worked on the automobile in his father's office at every opportunity. However, Teague's greatest marketing success was the Baby Brownie camera for the Eastman Kodak Company. The principle behind this camera was that such a product should be made as simple as possible for the amateur and dependable within fixed limits. Four million Baby Brownies were sold to the public at $1 each, the same price the original Brownie had sold for in 1900. By 1934 the Teague office was well established, with 14 clients across the country.

Russel Wright typifies a design position between the aforementioned marketing-oriented illustrators-turned-designers and the artisans of the Arts and Crafts movement. Beginning with a craft orientation, he developed (among other products) successful table accessories in various materials that could be produced with minimal investment in tooling expenses. Wright saw the humble products that serve the everyday needs of Americans as elegantly simple forms that express their material and method of manufacture in harmony with the purpose they were created to serve.

Industrial design had reached a comfortable maturity by the mid-1930s. Designers had demonstrated their ability to make manufactured products attractive to a reluctant public, and they were discovering a style that was uniquely American—a style that turned from the past to look hopefully to the future.

The sad economic realities at the opening of the 1930s had been challenged in New York by the soaring optimism of the neo-Gothic Chrysler building (1929), the sheer scale and dirigible mast of the Empire State Building (1930), and the modern classicism in the sculpture and architecture of the Rockefellers' Radio City complex. With the establishment of building codes that required a decrease in volume as building height increased, the skyscraper style emerged. The new

buildings' vertical thrust and disappearing mass as they reached for the clouds, enhanced by Hugh Ferriss's dreamy illustrations of modern castles in the air, became symbols of hope, and for a time the skyscrapers' form and decoration were emulated by everything from radios and scales to furniture and packaging.

However, the vertical motif of the skyscrapers was not dynamic enough for smaller objects. The horizontal lines of speeding vehicles were considered truer symbols of progress. "Speed lines" (usually three chromium-plated bands) became the distinguishing mark of the moment. They also proved to be a convenient and economically expedient way of applying a modern look to any product. Painted stripes could be substituted for metal bands if the occasion warranted it. "Straight lines," wrote Paul Frankl, "are typical of present-day directness." (35, 47)

Then, as monocoque aircraft competed for speed, altitude, and endurance and as airlines began to spin their webs across the United States and around the world, their smooth shapes began to dominate the form of products for the remainder of the decade. Just as the great ocean liners of the last two decades, with their clean white surfaces, portholes, ventilators, and metal tube railings, had influenced the international style of white surfaces, furniture and furnishings made of metal tubing, and round windows and mirrors, so now the aerodynamic form was to set the aesthetic direction.

The perfect aerodynamic form was believed to be a teardrop plowing through space with the round end forward. After all, fish as well as aircraft were shaped that way, so the teardrop was accepted as the ideal shape for all vehicles. Steamships and locomotives sought to disguise their embarrassment at being left behind by the airplane by donning streamlined shells. In this they were simply following one of the fundamental laws of the design ethic, by which a product that is nearing the end of its period in history takes on the form of its successor in order to stave off oblivion. Another law recognizes that, in the hierarchy of products just as in that of humans, the dominant class at any moment will set the behavioral pat-

Walter Dorwin Teague's 1932 Marmon 16 was one of the classic designs of the era. The replacement of the harsher geometric forms of the past with softer forms that were more suitable for stamped metal predicted the form of automobiles to come. Vanity Fair, 1932.

Teague's all-plastic 1934 version of the Brownie camera had vertical stripes in the skyscraper style. Pencil Points, September 1937.

The distinctive form of the Chrysler building (William Van Alen, architect), with its machinelike stainless steel ornamentation and its gargoyles, has made it a New York landmark. It is shown here between one of the towers of the Waldorf-Astoria hotel (right) and the RCA Victor building (left). Cervin Robinson.

The Empire State building
(Shreve, Lamb, and Har-
mon, architects) still
dominates midtown Man-
hattan, and its additive
masses established a
typeform for the American
skyscraper that has not
been equaled by any other
building. Hugh Ferriss
drawing; Architectural
Forum, June 1930.

*Rockefeller Center,
or Radio City. Thomas
Airviews.*

Donald Deskey's grand Art
Moderne interiors for the
Radio City Music Hall
have been compared to
those of the best European
designers. The main dif-
ference is that his have
been preserved as a
national monument while
theirs have long since
faded away. Anne Brown.

The new 1932 container
for Bon Ami cleaning
powder echoed the sky-
scraper form. It was pro-
moted as the smartest and
most convenient in the industry, and was intended
to be so good-looking that
it could be displayed in
the home. House and
Garden, *December 1933.*

BON AMI

A bookcase from 1930,
"rising up against the wall
like some building against
the sky" according to its
designer, Paul T. Frankl.
The piece was finished
in two tones of lacquer
and is set on a cabinet
base. Frankl, New
Dimensions.

Joseph Sinel designed this scale in the sky-scraper style for the International Ticket Scale Corporation. He recalls: ". . . I made a model from a Del Monte fruit case. . . . I took the model in to show the executives of the company and they immediately OKed it and they paid me a fee of $10,000 and royalties for twenty-five years." Sinel papers, California College of Arts and Crafts, Oakland.

In contrast to the vertical stripes on the earlier Brownie camera, Walter Dorwin Teague's 1936 Kodak Bantam Special was striped horizontally (some called this "streamlining"). Eastman Kodak Co.

Teague and his staff standardized Texaco's gasoline stations, with three parallel lines as the main ornament. Architectural Record, September 1937.

The Metropolitan Museum of Art invited several industrial designers to assist the architects who were doing the 1934 Contemporary Industrial Art exhibition. Lee Simonson worked with Raymond Loewy (shown here) to create this mockup of an ideal industrial designer's office. Note the three-line motif. Raymond Loewy.

In this 1932 photograph
the Pan American China
Clipper flies past the still-
incomplete Golden Gate
bridge on its inaugural trip
to the Orient. Pan Ameri-
can Airlines.

When Norman Bel Geddes
was asked to redesign the
interiors of the Clipper
airplane to provide long-
range comfort, including
sleeping and eating facilities, he created a system
that also took into account
maintenance and servic-
ing. Hoblitzelle Theatre
Arts Library, University
of Texas, Austin.

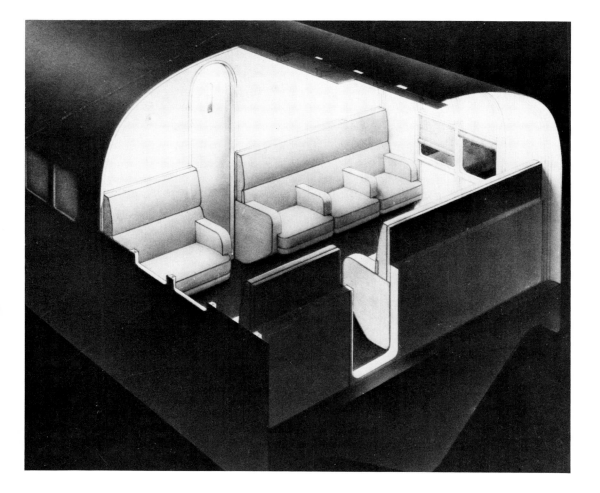

More than any other air-
craft, the Douglas DC-3
became the typeform for
1930s airliners. It went into
service in 1936 on Ameri-
can Airlines' Chicago –
New York route. This plane
set the character of aerodynamic form and thus
influenced many other
products. American
Airlines.

tern and cultural form of the subdominant class. Therefore it should not be a surprise that the streamline form came to dominate other products. (This transference of form has never been more evident than today, when the television screen has directed the shape of most contemporary products, from automobiles to wristwatches.)

Norman Bel Geddes secured patents in 1931 for his designs for a completely streamlined train that was quite reminiscent of Reverend Samuel Calthrop's patent of 1865. Although neither of these trains was ever built, the designs influenced the new articulated train that was built and operated by the Union Pacific Railroad. Within a short time all the railroads became aware that streamlining could help them counter the threats to the train's near-monopoly in overland travel from the airlines (which were drawing away first-class passengers), from the highways (which were being built with government funds), and from trucks (which were using the highways to haul the freight that used to be carried by the railroads).

The first fully streamlined train, the diesel-powered *Zephyr*, with its stainless-steel exterior and air-conditioned cars, was billed as a new type of train. The *Zephyr* was promoted with several barnstorming trips (one to help open the Century of Progress Exposition in Chicago) before it began full service in late 1934. By the end of 1935 every major railroad either was rebuilding or had rebuilt one or more of its passenger trains into a streamliner. These trains, with proud names like *Flying Yankee, Rebel, Comet,* and *Royal Blue,* captured the public imagination with their promises of speed. The two major eastern railroads, the Pennsylvania and the New York Central, dominated the field of streamliners.

Raymond Loewy was employed by the Pennsylvania Railroad in 1934 to produce a streamlined shell for its GG-1 electric locomotive. Borrowing an idea from automotive production lines, he recommended that the traditional rivet construction be replaced by welding to achieve a smoother shell. Loewy's second Pennsylvania Railroad assignment, to provide a shell for the PRR steam locomotives (beginning with Engine 3768), was considered particularly successful

because, rather than disguising their boilers, he emphasized their form and promise of power.

Henry Dreyfuss' first railroad assignment, from the New York Central, was to design the luxury streamlined train *Mercury* (whose locomotive was to be a remodeled 1916 machine). This was a complete job, including not only a new form for the locomotive but also an entirely new layout for the car interiors, down to hardware, furnishings, and tableware. Dreyfuss used spotlights to emphasize the locomotive's powerful driving wheels—a device not unlike the illumination of vertical stabilizers and rudders on modern commercial airliners. However handsome and thorough his concept for the locomotive was, it still looked somewhat out of tune with the more familiar boiler-dominated typeform.

The era of the streamliners came to a climax in 1938 when the New York Central's *Twentieth Century Limited* and the Pennsylvania Railroad's *Broadway Limited* were introduced with great fanfare on the same day in New York and Chicago. For a special surcharge, their first-class passengers were treated to the ultimate in modern design, luxury, and convenience. The *Twentieth Century Limited,* designed by Henry Dreyfuss, was essentially an upgraded version of his earlier *Mercury.* However, the locomotive had been redesigned more along the powerful lines of Loewy's PRR engines. The *Broadway Limited,* designed by Raymond Loewy in collaboration with Paul Cret, head of architecture at the University of Pennsylvania, maintained the bold locomotive forms of the PRR while borrowing liberally from the interior design of the *Mercury.* Until the American entry into World War II, these two trains competed head to head for luxury-minded passengers on the lucrative Chicago–New York run.

Although aerodynamics had been considered appropriate for automobiles at the turn of the century, it was not until the 1930s that the technology for stretching larger sheets of thin steel into complex shapes and welding them became economically feasible for production automobiles. Both European and American designers and

There was a strong feeling in the 1930s that, had the railroads wanted to, they could have competed effectively in speed and service against the fledgling airlines. Norman Bel Geddes proposed and patented this completely streamlined train that would have tested the premise had it been built. Hoblitzelle Theatre Arts Library, University of Texas, Austin.

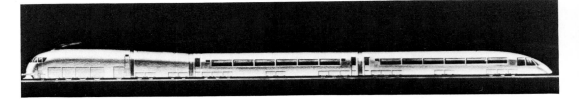

"Before" and "after"
photographs illustrating
Raymond Loewy's 1934
redesign of Pennsylvania
Railroad's electric loco-
motive, the GG-1. The
all-welded shell and the
smoother contours in-
creased speed and per-
formance. Raymond
Loewy.

Loewy designed the streamlined shell for the Pennsylvania Railroad's Engine 3768 after extensive wind-tunnel tests had shown that the engine's wind resistance could be lowered by one-third. The locomotive was used to pull the Broadway Limited *for two years, then was put on display at the 1939 New York World's Fair.* Architectural Forum, September 1938.

The interiors of the Broadway Limited *were designed by Raymond Loewy and Paul Cret. The interior view of the Bar Lounge car shows the thoroughness and cleanness of their design.* Raymond Loewy.

The 1938 Twentieth Century Limited *was designed by Henry Dreyfuss and his staff after their work on the successful* Mercury. *In effect the* Century *was a new* Mercury *in a much more luxurious form. Its engine remains one of the most distinctive typeforms of the era.* Architectural Forum, *September 1938.*

The dining room of the Twentieth Century Limited *illustrates the elegant modern style upon which its fame was based. Dreyfuss used optical illusions to make the cars appear wider. The walls had cork to deaden sound, and the tables were cantilevered.* Architectural Forum, *September 1938.*

engineers experimented with various smooth-shaped bodies, usually for rear-engined vehicles. The Moglia and Claveau from France, Sir Dennistown Burney's teardrop, Edmund Rumpler's Tropferwagen, William Stout's Scarab, and R. Buckminster Fuller's Dymaxion car were all conceived as airfoils or teardrops in order to decrease wind resistance, thereby improving speed and performance and reducing fuel consumption. None of these was ever manufactured successfully, but they helped to prepare public opinion for the automobiles that were to follow. In 1931 the Society of Automotive Engineers concluded that the teardrop was the ultimate form for automobiles and predicted that before long manufacturers would offer rear-engined teardrop automobiles to the public.

Although industrial designers did not have an important role to play in the technological development of aerodynamic forms for automobiles, several of them were contracted by automobile manufacturers to develop design concepts that would explore the influence that the trend toward streamlining would have on their vehicles. Norman Bel Geddes for Graham-Paige, Raymond Loewy for Hupmobile, and Walter Dorwin Teague for Marmon recommended a softening of details and the use of slanted surfaces to reduce air pressure under speed. In general, their designs produced a much more handsome effect without a fundamental change in form. Although Loewy went so far as to build a full-scale model in an effort to convince Hupmobile management to manufacture the automobile, only Teague's Marmon 16 was ever manufactured, and then only for three years. Nevertheless, the industrial designers were attracted to the possibilities that streamlining offered. Norman Bel Geddes designed, built models of, and secured patents for a teardrop automobile, a bus, a yacht, an ocean liner, and an airplane. Raymond Loewy also obtained patents in 1928 for an automobile with a strong streamline profile. In addition Loewy was commissioned to apply the characteristic sweep and curved nose of the streamline form to the boxy sides of buses for the Greyhound company. He was also able to carry through the handsome streamlining of the *Princess Anne* ferryboat.

Only one truly streamlined automobile by an American company was ever carried through to production. In 1934, the Chrysler Corporation, after several years of wind-tunnel studies under the direction of engineer Carl Breer (which proved that conventional automobiles of the era encountered less resistance when moving backward than they did when moving forward), introduced the Airflow automobile. It was as faithful as possible to the experimental aerodynamic form, with recessed headlights in a smooth front and a slanted back. In its first year of production, 11,000 Airflows were sold. In the following year 8,000 were built, with a protruding V-shaped grill designed by Ray Dietrich to overcome public resistance against the flat front. In 1936 an even smaller number (4,000) were manufactured, with other innovations, including an all-steel "turret" top and a fully enclosed trunk. Despite the fact that the DeSoto version had won two consecutive grand prizes in the Monte Carlo Concours d'Elegance for its aerodynamic styling, and despite the wide publicity its daring form and innovations had attracted (Norman Bel Geddes even appeared in advertisements, extolling its virtues), the Airflow was not a commercial success and was dropped by Chrysler in 1937.

The influence of the Airflow on other automobiles was unmistakable. The V front and the slant back became standard in the industry, and by 1939 the formal differences between one automobile and another were so slight that graphic identification had to be used to distinguish them.

The Lincoln Zephyr managed to capture the best expression of the style, but the Airflow is still recalled with respect as the first attempt by an American automobile manufacturer to break away from the typeform of the carriage. Although it was primarily a commercially motivated experiment in form, the Airflow illustrates that innovation in design carries a threat of failure that is commensurate with its promise of gain. Failure in the marketplace is the price the design ethic extracts from those who dare to step too far ahead of evolutionary development. Raymond Loewy has followed the motto that one should strive for a design that is "MAYA"—the most advanced yet acceptable.

R. Buckminster Fuller's
Dymaxion car was shown
and demonstrated at the
Chicago fair. Fuller is
shown here with two
passengers. R. Buck-
minster Fuller.

STREAMLINED BUS: DRAWING.   Courtesy, Texas University.

Bel Geddes's streamlined double-deck motor coach would have reduced wind resistance to the same degree that it increased passenger comfort and convenience. Hoblitzelle Theatre Arts Library, University of Texas, Austin.

Bel Geddes designed this seagoing yacht for Axel Wenner-Gren. Although it seems like pure fantasy in retrospect, every detail was based on an environmental or operational conviction. Hoblitzelle Theatre Arts Library, University of Texas, Austin.

With Otto Kuhler, Norman Bel Geddes dreamed up Super Airliner 4 in 1929. It was to carry 450 passengers and 115 crew members at a cruising speed of 100 mph, driven by 37 propellers. Hoblitzelle Theatre Arts Library, University of Texas, Austin.

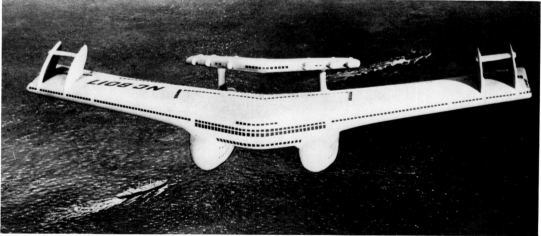

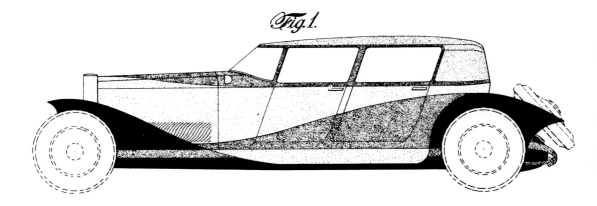

Fig.1.

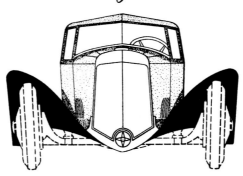

Fig.2.

Raymond Loewy's design patent for "combined automobile body, hood, and fenders" shows an interest in achieving a unified streamline effect—no matter what the underlying body form was. U.S. Patent 79,147, August 6, 1929.

In 1933, Loewy redesigned a ferryboat for the Pennsylvania Railroad. Put into service in 1936 and rechristened the Princess Anne, it is generally considered the most successful streamlined vehicle of the time. Raymond Loewy.

The Chrysler Airflow was
the first production automobile to be designed
according to aerodynamic
principles. It made a clean
break with earlier designs
and established an aes-
thetic base for the Lincoln
Zephyr and other cars
to come. Chrysler
Corporation.

Once a viable design form
is established, all of the
competitors adopt it without hesitation. The louder
they proclaim their dif-
ference the more they are
the same. Mark Waterman.

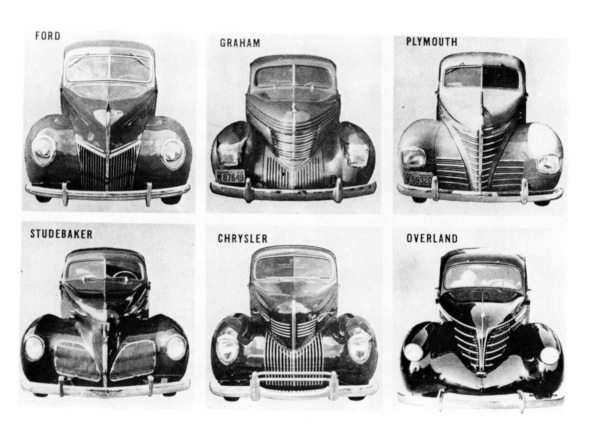

FORD

GRAHAM

PLYMOUTH

STUDEBAKER

CHRYSLER

OVERLAND

The Lincoln Zephyr, built
for Ford by the Briggs
Manufacturing Company
in 1935, was called the
"first successfully de-
signed streamlined car
in America" by the Mu-
seum of Modern Art. The
front and the rear were
shaped to a "V." Fortune,
May 1936.

The Lincoln Zephyr trade-
mark had the teardrop
shape that was consid-
ered aerodynamic perfec-
tion. Fortune, May 1936.

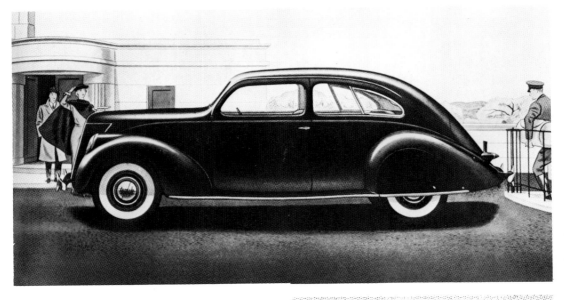

If other manufacturers had followed the lead of the Airflow a real change in automotive form may have occurred. They did not. However, in 1938 the cornerstone was laid at Wolfsburg for the factory that would produce the German equivalent of the Airflow, Dr. Ferdinand Porsche's Volkswagen. Over 40 years later the "Beetle" still appealed to the public and was still being manufactured in several countries, and its total sales had surpassed the Model T's 15 million.

The public's interest in streamlining offered designers the rationale they needed to make it an aesthetic base for mass-produced objects. For a time it was applied to every product possible, resulting in what some have called the "streamlined era," and in the process it seems to have fixed in the public mind the notion that industrial designers are "streamliners." For a few years the characteristic streamline was accepted as the ideal form for many American manufactured products, much to the dismay of some observers. Edgar Kaufmann, Jr., wrote sarcastically that "the teardrop swelled, divided and multiplied, became garnished with ribbons of chrome and elevated on an altar of sales, while statistical Magnificats were sung in its honor." (146, 89) Laszlo Moholy-Nagy saw streamlining as "superficial styling" to which industrial designers had succumbed under pressure from salesmen, yet he later admitted that the new form increased the strength of a product's shell and made it easier to manufacture.

Henry Dreyfuss came closest to recognizing the real contribution of streamlining to products. Although he was concerned about those designers who accepted half-truths about streamlining and applied it indiscriminately to pencil sharpeners, fountain pens, and the like, he believed that it had benefited American products. "The designer," he wrote, "learned a great deal about clean, graceful design. He learned to junk useless protuberances and ugly corners." Dreyfuss suggested that the style should be called "cleanlining instead of streamlining." (27, 75) And Harold Van Doren, like most American designers of the time, saw no harm in making manufactured products conform to the full-flowing forms of modern aircraft, in place of the harsh angularity

of the modernists or the cold intelligence of the functionalists. At the very least, streamlining was the first new and uniquely American approach to form that the public could associate with progress and a better life.

In retrospect, streamlining seems to have offered more advantages than disadvantages. Manufacturers accepted it as a practical means of simplifying production. The shell could be disassociated from its structural skeleton, its mechanical organs, and its energy and control systems, thus allowing more economical production and increased efficiency and dependability. At the same time the exterior of the shell could be better adapted to the physical and psychological needs and desires of its owners as well as to the demands of its distribution and marketing environment.

Manufacturers now recognized that industrial designers seemed to be able to anticipate the special intersection of onrushing technology and volatile public preference. Therefore, they sought them out for their magic touch, and the designers—in a sense, outsiders to industry—were accepted as monitors of public taste. By adding technological understanding to their aesthetic sensitivity they proved that they were able to close the gap between humans and the products that were being manufactured to serve them. Half a century later the concept of employing designers to help industry increase sales would be held suspect by the humanists, but during the Depression if a designer could help a manufacturer escape bankruptcy and preserve the jobs of his employees he was welcomed as a savior. As a result, companies hastened to establish their own internal design groups or to retain independent designers as consultants.

Harold Van Doren and his associates designed this "Tot Bike" for the American National company in the streamline style. Architectural Forum, *October 1940*.

It is entirely natural that the "airstream" concept should have been applied to the electric fan. This fan is by Robert Heller for the A. C. Gilbert company (1938). Architectural Forum, October 1940.

Raymond Loewy's whimsical 1934 application of the streamlined form to a pencil sharpener was much criticized. Raymond Loewy.

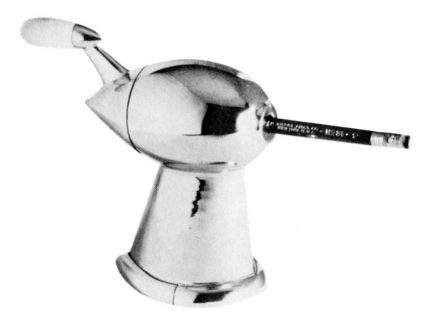

From "Cleanlining" to Accountability

We see that we are not building big or little gadgets—we are building an environment. And we designers have to work also with the scientists, engineers, technologists, sociologists and economists who have part in this reconstruction. Can we get enough of this new world strongly and fairly built in time?

Walter Dorwin Teague, 1940 (95)

The public's interest in design was stimulated in the mid-1930s when the government authorized a Federal Art Project to establish an Index of American Design. The idea was proposed by Ruth Reeves, a textile designer and painter, to the New York Public Library and then carried to Washington for endorsement. With the noted American historian Constance Rourke as national editor and Ruth Reeves as the first national coordinator, the project ran from late 1935 until the United States entered World War II, covering some 35 states and employing an average of 300 artists at a time. It produced over 17,000 carefully detailed illustrations of American decorative and industrial arts as well as vernacular products dating from the earliest days of the colonies until the end of the nineteenth century. Although the project's immediate purpose was to provide employment for commercial artists during the Depression, its end value was to record and thereby honor the indigenous arts and industries of the Americans. This important survey of the objects of everyday life served to dignify them as well as to preserve them. Constance Rourke saw an even deeper value in the collection (which is now owned by the National Gallery): "If the materials of the Index can be widely seen they should offer an education of the eye, particularly for young people, which may result in the development of taste and a genuine consciousness of our rich national heritage." (45, I, xxvii)

The widespread interest in industrial design was forcing major schools and colleges to consider adding it as a subject. The primary question was how this was to be accomplished within a rigid, self-serving academic system. The new discipline was neither rational enough to be acceptable to engineering schools nor noble enough to be welcomed by architecture. Furthermore, well-established art schools—particularly those with strong arts and crafts programs, such as the University of Cincinnati or Cooper Union—either were too busy to take in the young orphan profession or else had already expanded their class assignments to include the design of manufactured products in overlapping areas, such as radio cabinets, lighting fixtures, or tableware. However, there was a prevalent belief that the arts and crafts schools lacked the vision that would enable them to submit to the existing capabilities of industries or to subscribe their work to the daily needs of any particular segment of the buying public. Once again, it appears, the stimulus to break with established habit had to come from abroad—this time not from France, but from an unexpected source in Germany.

When Walter Gropius was invited to establish a school at Weimar in 1919, he accepted only on the condition that he be allowed to carry out the idea of his predecessor, Henri Van de Velde, that the fine and applied arts be combined in an academic experiment to demonstrate their fundamental unity. This bold venture attracted a distinguished faculty from all over Europe and students of all ages, incomes, and political persuasions.

Gropius's particular goal for Das Staatliches Bauhaus Weimar was to establish a "consulting art center for industry and the trades" that would break down the barriers between artists and craftsmen by means of an innovative preliminary course of six months by which he hoped to cleanse the students of their previous training, release their intuition and sensitivity, and encourage them to experiment with old and new materials. (8, 12) This course's methodology was a "learning by doing" experience that was more than casually related to training in the crafts. To some observers, in fact, the course was little more than an extension of the English and Viennese Arts and Crafts movements without, as Gropius pointed out in defense, their affection for romanticism or their commitment to "l'art pour l'art." (8, 21)

The Index of American
Design represented the
work of over a thousand
illustrators and produced
over 22,000 plates cov-
ering the arts and crafts
and everyday products of
the Americans. This photograph shows a Works
Progress Administration
artist at work. National
Archives.

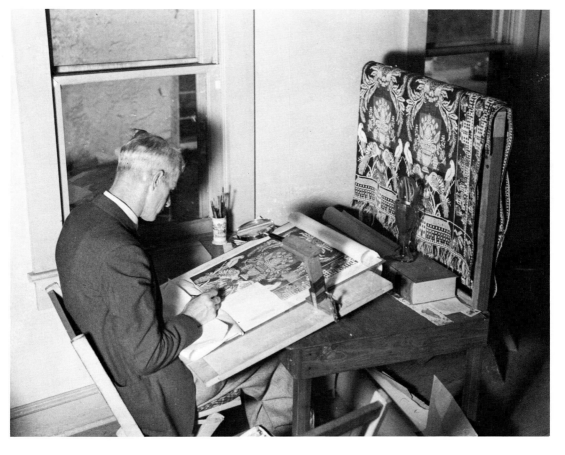

By 1923, as the exuberant behavior of those involved with the free mixture of the arts in the Bauhaus began to wear thin its welcome in Weimar, Gropius found it prudent to accept an invitation to move the entire school to Dessau. As the move was being made, the programs modified, and the faculty reassigned or replaced, it became evident that the philosophic emphasis of the school was shifting from Expressionism and the fine arts toward the conflict between rationalism and formalism in design. Its previous preoccupation with handicraft methods as a prelude to design for machine production was now to be concentrated, as Gropius remembered some ten years later, on averting "man's enslavement by the machine by giving its product a content of significance and reality." (212) Gropius saw rationalism as only a purifying force and not a cardinal principle, and warned that formalism was merely a fashion in modern art. Nevertheless, a mannered Bauhaus character began to appear by which products were styled to give an illusion of industrially made things whose geometric form and visibly mechanistic construction were their ornament. In this context it is interesting to remember that, some 50 years earlier, machine methods were used to manufacture products that appeared to have been made by hand. And now the Bauhaus workshops were using handicraft methods to produce objects that pretended to have been made by machines. Reyner Banham quotes Laszlo Moholy-Nagy's criticism of Wilhelm Wagenfeldt for having changed cylindrical milk jugs into drop-shaped ones: ". . . how can you betray the Bauhaus like this? We have always fought for simple basic shapes, cylinder, cube, cone and now you are making a soft form which is dead against all we have been after." (5, 282)

By the time that the new building was finished and the Bauhaus was in full operation again, Walter Gropius realized that it had carried with it from Weimar the germs of its own dissolution. On one hand he regretted the formalism of the Bauhaus style as a "confession of failure and a return to the very stagnation, that devitalizing inertia, to combat which [he] had called it into being;" on the other hand he deplored what he called "spurious phrases like 'functionalism' (*die*

neue Sachlichkeit) and 'fitness for purpose equals beauty' " championed by the socialists in the Bauhaus. (212) As a result Gropius gave up the directorship of the school to Hannes Meyer, a Swiss communist, stating that its intellectual objectives had been attained, and returned to Berlin to devote his time to architecture. With him went Marcel Breuer, Herbert Bayer, and Laszlo Moholy-Nagy, leaving the school entirely in control of the socialists.

Hannes Meyer was obsessed by the extreme practicality of functionalism and instituted severe academic rules. His favor toward communist students so offended the authorities that in 1930 he was forced to resign. His place was taken by Ludwig Mies van der Rohe, the Berlin architect who was the director of the Deutscher Werkbund. Mies promptly closed the school for a month and then reopened it, without the communists, as a school only for those interested in pure architecture. The school was finally closed permanently by the National Socialist government in 1933. The political side of the Bauhaus story takes a somewhat different tack. It suggests that the original breakup of the Bauhaus was less a matter of conflict with the National Socialists than it was an ideological disagreement between the presumed formalism of Walter Gropius and his adherents and the functionalism of Hannes Meyer and his fellow-travelers. When the formalists abandoned Dessau, Meyer took over until the Nazis ran him and his group off to Moscow. Mies's one year as director of the Bauhaus after it was moved to Berlin was merely a postscript to the history of this important shrine of design education.

Over barely a decade the Bauhaus had been seeded at Weimar by the best design minds of Europe, had been brought to full bloom at Dessau, then had withered in the conflict between formalism and functionalism and finally been uprooted and trampled in the political upheavals. However, the dynamism of its principles and the plight of its adherents captured the imagination and sympathetic attention of the Western countries, and many of its most illustrious faculty members and students found a home in the United States.

Anni and Josef Albers, who had been with the Bauhaus from the first days at Dessau, were the first to come. With the closing of the school in Berlin they emigrated to take a position at Black Mountain College in North Carolina. Walter Gropius and Marcel Breuer moved to London until 1937, when both came to the United States, Gropius to head and Breuer to teach in the school of architecture at Harvard University. Walter Baermann came over to establish a program in design at the California Institute of Technology, and Hin Bredendieck to teach at the New Bauhaus in Chicago when it was established.

Laszlo Moholy-Nagy, who had gone to England with Gropius, came to the United States in 1937 on Gropius's recommendation to head the New Bauhaus in Chicago, which was to be funded by that city's Art and Industries Association. It was hoped that the New Bauhaus would embody the original Bauhaus's principles and traditions of integrating art, science, and technics. Despite a successful first year, the school was forced to close because its sponsors were unable to raise the additional funds that were necessary. By the following summer, however, the school was reborn under the name of the School of Design. Herbert Bayer came to the United States in 1938 to work with Walter and Ise Gropius on the Museum of Modern Art's retrospective exhibition "Bauhaus, 1919–1928." This exhibition, directed by curator of architecture and industrial design John McAndrew, stimulated considerable contention among critics as to whether the Bauhaus was dead or whether it was indeed being reborn in the United States.

Gillo Dorfles has evaluated the transplantation of the Bauhaus group to the United States as not "destined to transform suddenly the quality of American industrial design and bring it closer to European taste and procedures." Wrote Dorfles:

The United States, with its highly advanced industrialization and important contributions already made to modern architecture and industrial design, was not greatly affected by the Utopian zeal of the German artists. In the years between the two world wars America had already seen the large-scale development of styling; considerable

attention had been given to the external appearance of products. Large design-consulting studios for industry already existed in the United States and appliances, airplanes and automobiles designed by these studios anticipated in their extensive styling and mechanical perfection the products that were to appear in Europe after World War II. (208)

The fact is that the Bauhaus expatriates, despite the sympathetic attention that was paid to them by a select audience, did not find their way into American industry. However, they were eminently successful in bringing a fresh approach to the foundation courses of American design schools.

E. M. Benson had proposed in a perceptive 1934 essay in the *American Magazine of Art* that the United States should establish a school that would pick up the cause of design for industry where the Bauhaus had been forced to leave it. Benson recommended that an American Bauhaus be established that would follow a parallel academic path by dedicating the first year of the design student's education to the theory and practices of materials and manufactures. After that, said Benson, the students should be required to branch off into a specialized field. He also insisted that the executive board of the school should include sociologists and manufacturers. In his interest in adding a sociological component to the designer's education Benson reflected the Bauhaus principle that technology should be made subservient to personal and societal values. Benson's article was accompanied by a photograph of the Bauhaus at Dessau and by illustrations of plans prepared by Frederick J. Kiesler, an innovative architect and designer who had emigrated from Austria some years earlier, for such an "Institute of Art and Industrial Design."

In 1935 the Works Progress Administration expanded its interest in design to include design education by providing a grant for the establishment of just such a school in New York—to be called the Design Laboratory—for students who could not afford private schools for training in design and fine arts. Gilbert Rohde was appointed director of the school, and Walter Dorwin Teague, Henry Dreyfuss, and other designers were on the board. Rohde claimed that the cur-

Gilbert Rohde. Architectural Forum, *January 1936.*

riculum was patterned after the Bauhaus in
Germany yet modified somewhat to coordinate
training in aesthetics and product design with
studies in machine fabrication and merchandising. A cut in WPA appropriations forced the
Design Laboratory to close within a year as a
federally supported educational institution. (The
school managed to continue for a short time on
its own on a charter from the New York State
Board of Regents, with a new name, the Laboratory School of Industrial Design.) The abandon-
ment of the first and only example of federal
interest in the education of industrial designers
left the United States the only major country
whose national government does not support
industrial design, either in education or as an
activity related to the development of trade and
industry.

After decades of abortive attempts to establish
academic programs for training industrial designers, resistance disappeared quickly in the 1930s
when several institutions of higher learning introduced curricula in industrial design under the
tutelage of respected designers. Their intent was
not only to teach the fundamental skills of the
new profession but also to establish a foundation
of knowledge that would analyze human needs,
consider alternatives, and recommend appropriate solutions. The first degree-granting program in industrial design was established in 1935
at the Carnegie Institute of Technology in Pittsburgh. Donald Dohner, who had earlier taught
design at Carnegie, was succeeded by Peter
Muller-Munk, Robert Lepper, and Alexander
Kostellow. New programs were also started at
Brooklyn's Pratt Institute (by Donald Dohner, with
Gordon Lippincott and later Alexander Kostellow),
at New York University (under Donald Deskey),
and at Columbia University (under Frederick
Kiesler). The Universities of Syracuse, Cincinnati,
and Illinois, the Art Schools in Chicago, Cleveland, Philadelphia, Dayton, and Milwaukee, and
the Rhode Island School of Design evolved industrial design programs from established
courses in the applied arts. At one time or another virtually all of the first generation of industrial designers organized and taught courses in
industrial design or else served on the advisory
board of one school or another. Thus these men,

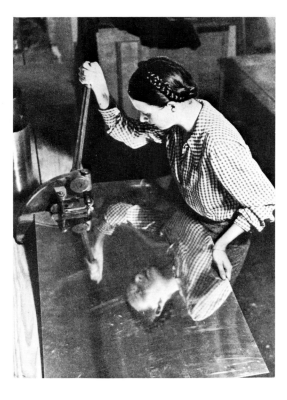

The Design Laboratory's metal workshop provided design students with an opportunity to become familiar with modern materials—in this case, aluminum. American Magazine of Art, *October 1936.*

who were themselves not formally educated in the field, drew from their experience the knowledge and wisdom to establish a practical base for industrial design education in the United States.

In an article published in 1938 in the magazine *PM,* Grace Alexandra Young noted that Europeans could not understand the assurance with which Americans plunged into jobs without years of study and apprenticeship. Europeans, she observed, preferred to organize schools and write books before they began to practice, whereas Americans preferred to do things first and then to "form an organized philosophy from the results." (203, 26) As if in confirmation of Aristotle's dictum that art runs ahead of its theory, the American way seemed to put practice before academic theory. If it had been otherwise the United States might still be dependent upon Europe for design. Instead, once given an opportunity, the Americans pressed for the introduction of design into virtually every industry without dependence upon prior theory. They shed the burden of the past and quickly learned how to make their way across the no-man's-land of commerce and industry without waiting for academic approbation or philosophic forbearance. Thus, they broke through on one front after another to take advantage of every chance to serve public expectations. All of this is in harmony with the unique ability of the Americans to react to the threat or opportunity of the moment, sustained by their instinctive faith in technology and its promise for the future. Americans believe in the ability of a free people to select the good from the bad and to tolerate social controls only when they become demonstrably necessary. They have no taste for autocratic rule from a cultural, social, economic, or political aristocracy. Though they may be impulsive and even brash at times, there is an inherent drive at work that presses the whole toward human rather than mechanical values.

In its creative vigor, American industrial design was often impatient with industry's cautious movement toward the better environment that was promised by a rapidly advancing technology. Many designers used their free-wheeling talents to conceive and propose to the public advanced concepts that drew attention to new possibilities and generated a public restlessness and a desire to move forward more quickly in the future. The automobile manufacturers used designers' "dream cars" to attract attention at auto shows. The general public was so flattered by such appeals to its aspirations that "dream" products and houses and kitchens of "tomorrow" became part of a "blue-sky" design ethic that offered a tantalizing glimpse of the future without a firm commitment that it would be delivered. It was generally expected that trade shows and expositions would include futuristic ideas. Even when farfetched such designs were, at the very least, entertaining and exciting; when they were closer to reality they managed to pull a product forward and upward by clearing the way for change.

Industrial design as a truly modern generalist profession had found an important place for itself among art, engineering, and business, though it had not as yet established a firm academic base. As the profession's glamor attracted a host of less qualified opportunists, its glittering reputation began to tarnish. Apocryphal stories began to be circulated about exorbitant fees and runaway royalties for designs that did not always come up to their promises. An awakening conscience about the integrity of manufacturers and the impact of their products on the economic and physical environment placed blame on designers for artificially limiting the effective life spans of some products and violating the trust of the public. Some of the preeminent designers of the time found it necessary to warn against the notion that design was a universal panacea for industrial ills and to disclaim exaggerated stories of success. Henry Dreyfuss emphasized his conviction that the spectacular and the sensational had no place in most products and suggested that designers should consider themselves accountable to consumers. As a result, conscientious and experienced designers came to the conclusion that they must organize to monitor their own behavior and to set themselves apart from less responsible usurpers of the unprotected title of industrial designer.

In 1938 one group of designers met in Chicago at the instigation of Lawrence Whiting, head of the American Furniture Mart, Alfred Auerbach,

editor of *Retailing Daily,* and Richard F. Bach of the Metropolitan Museum to form the American Designers Institute (ADI), with John Vassos as president, and to subscribe to a code of ethics written to protect their clients and the public. There were some 45 founding members, most of them connected in one way or another with the home furnishings field. Later, after President Edward Wormley of the Chicago chapter and other designers rebelled against the transparent paternalism of non-designer members from marketing and industry, the organization was rechartered in New York State and its name was changed to the Industrial Designers Institute (IDI) in order to attract a broader group of professional designers and to deemphasize its association with craftsmen. The IDI also absorbed the Chicago Society of Industrial Designers, of which Dave Chapman had been a founder.

The second major organization, the Society of Industrial Designers (SID), was established in 1944 in New York City after the courts ruled in a 1941 Unincorporated Business Tax suit versus Walter Dorwin Teague (who was supported by other prominent industrial designers) that Teague's primary activity was providing a service to the public and that, therefore, he was not liable for the tax. This has been taken since then as *prima facie* evidence that industrial designers are professionals. The fourteen founding SID members were all practicing industrial designers: Egmont Arens, Donald Deskey, Norman Bel Geddes, Lurelle Guild, Raymond Loewy, Ray Patten, Joseph Platt, John Gordon Rideout, George Sakier, Jo Sinel, Brooks Stevens, Walter Dorwin Teague, Harold Van Doren, and Russel Wright. In 1955 the SID changed its name to the American Society of Industrial Designers (ASID). In 1965 the IDI and the ASID merged with the younger Industrial Designers Education Association (IDEA) into a single professional voice for Industrial Design in the United States, the Industrial Designers Society of America (IDSA).

The successful establishment of the industrial design profession in the United States was the natural result of a general acceptance of the principle that progress and happiness are best ensured when better products are manufactured at a cost that is within the reach of everyone. Many designers shared Norman Bel Geddes's convictions that "design in machines . . . shall improve working conditions by eliminating drudgery" and "design in all objects of daily use shall make them economical, durable, convenient, congenial to every one." (37, 5)

The industrial designer was aware that every mass-produced product was in a constant state of evolution as it sought to keep up with changing personal and societal patterns of living as well as the inexorable advance of technology. He was certainly concerned about the meaning and value of his design recommendations to the consuming public and their effect on the reputation and economic condition of his client. And he was familiar with the so-called canons of machine art that were championed by cultural philosophers: geometry because of the mathematical behavior of mechanical devices, precision because of the need for absolute control of production technology, simplicity because of the inexorable pressure to refine the product at the same time as its performance is improved, and economy as the struggle to bring the cost of a product down to the level that constitutes its broadest market. These elements had deeper implications for the industrial designer than they appeared to have for the industrial stylist and the aesthetician. Geometry (or, better still, the trilogy of forms that Walter Gropius praised as the housekeeping of the mind—the square, the circle, and the triangle) implied a simplistic obeisance to these shapes without acknowledgment that, as in nature, the form of a product must be an expression of the inner nature and outward function of the product. To the uninitiated, precision implies sharp edges, smooth surfaces, and perfect fit, without the consideration that products conceived for mass production must allow tolerances and clearances that are essential for effective production and for expansion and contraction in manufacture and use. Simplicity to the industrial designer implies that forms must allow for easy and convenient assembly and maintenance and that reveals, setbacks, and textures must be employed to absorb assembly variation and wear. And economy

that reduces the manufacturing cost of the product must do so at no risk to the consumer's best interest.

At this point a schism began to develop between the stylists and the designers. The more articulate stylists of the period called for a "style of reason" that would appeal to intelligence rather than romanticism. (35, 31) Paradoxically, they were preoccupied with the visualization of reason rather than its actuality—with the creation of forms that appeared to have been made by machines. According to Percy Seitlin, "the industrial designer tackles his problem from the inside out. His design is based on function. . . . He is concerned with evolving a product that is honest, beautiful and of improved usefulness. The industrial stylist, on the other hand, is little concerned with the inside but very interested in the outside. He is a designer of shells and packages. His work is frequently more sensitive than the industrial designer's whose means are more subtle by comparison." (197)

Corporate hierarchies created another problem: Whereas more independent consulting offices insisted that they offered a complete design service for their clients, it was apparent that in-house designers were constrained by their position in the corporate table of organization. For the most part, they were assigned to either the marketing staff (where they were usually expected to serve the market forecast) or the engineering staff (where they were usually subject to the production and performance program). However, in some cases, as at General Motors in 1937, industrial styling was elevated to staff level. In recognition of the importance of styling, GM made its chief stylist Harley Earl a vice-president in 1938.

The middle road between styling and design is best represented by companies like Sears, Roebuck, which (after it employed John Hauser in 1935 to replace Jack Morgan, who had resigned to go into private practice) expanded its design activity to a full department of industrial design. Under Hauser the department ranged freely between design and styling in its effort to establish and control both the formal and functional values of the products that this great merchandising organization commissioned from manufacturers.

At the most challenging level of professional practice, perhaps, industrial designers began to reach beyond the narrower problems of solving either formal or functional assignments to the innovative development of products or product families that opened new marketing areas for manufacturers and offered new products and services to the consumer. Some of the most successful ventures in this area were achieved by industrial designers acting as designer-craftsmen to produce the prototype of a new product and then to adapt it to production in close collaboration with a manufacturer. W. Archibald Welden, Donald Deskey, Russel Wright, Gilbert Rohde, and (somewhat later) Eero Saarinen, Charles Eames, George Nelson, and Don Wallance are representative of this entrepreneurial approach to product development.

After the Kantack company closed in 1933 because the Depression had lessened the demand for architectural and gift metalwares, W. Archibald Welden was employed as an independent designer by the Revere Copper and Brass Company to develop a series of cooking wares. His exhaustive research and his meticulous concern for functional and manufacturing detail resulted in a line of stainless steel and copper products that is still being manufactured. Over the years this Revere Ware has achieved the status of an American typeform. Often imitated and even copied outright, it remains the leading line of cooking wares and is still being shown in advanced designs for the "kitchen of tomorrow."

In the late 1930s Donald Deskey, disturbed by the wild grain texture that resulted when Douglas fir logs were peeled to make the plywood that he wished to use as paneling on his sport cabin, developed a striated texture that he called "Weld-Tex." The process consisted of taming the surface of plywood with a random-width but straight-cut pattern of grooves. The new texture gained instant popularity, and after Deskey sold the rights to the texture to the United States Plywood Company it was used for well over a decade as a modern finishing material.

Russel Wright may have been the most versatile entrepreneurial designer of the 1930s. Early in the decade he had developed and introduced to market a line of "spun" aluminum giftwares that were available at a moderate price. Before this Wright had been employed by Edison Laboratories to develop a cocktail shaker with a coat of chromium applied by the electro-deposition process. By 1935 his previous experience (with the Wurlitzer and Capehart companies) with pianos, radios, and the first jukebox led him to develop a line of straightforward furniture made of solid bleached maple that he called "blonde." With its soft rounded "cushion" edges it projected a simple handcrafted look that struck a note between the Swedish and Early American furniture styles. The final line of some 50 pieces was manufactured and successfully introduced in 1935 by the Conant-Ball Company in Grand Rapids, Michigan. In 1936 Wright developed a unique, somewhat exotic chair in mahogany for the boardroom of the Museum of Modern Art.

In 1938 Wright and his wife Mary tackled a new material by developing a line of ceramic dinnerware that combined their interpretation of the fluid forms of streamlining with a fresh approach to functional design in soft colors and textured glazes. They invested their own funds in producing the original models in plaster and the molds from which the pieces could be manufactured. When they could not find a manufacturer that was willing to take a chance on their designs, they located a pottery in Steubenville, Ohio, that had been closed as bankrupt and convinced the authorities to reopen the plant to manufacture their wares. The American Modern line, as it was named, was introduced in 1939 and included in the Museum of Modern Arts Annual Exhibition of Useful Products Under $10 in January 1940. Deliveries did not begin until the fall of 1940 because of manufacturing difficulties. From that time until 1961, when production was stopped, the line earned over $1 million in royalties for the Wrights, and today American Modern is a collector's item.

Coincidentally, Russel Wright organized a nationwide consortium of over 100 outstanding industrial designers and artist-craftsmen in an

This rendering of the original Revere Ware saucepan, designed by W. Archibald Welden, shows the simple, distinctive form that has made it an American classic. Revere Copper and Brass Co.

The expanded Revere line of cookware, which combined the heat conductivity of copper with the cleanliness of stainless steel. Revere Copper and Brass Co.

This "blonde" maple line of furniture, known as "American Modern," was designed by Russel Wright and manufactured by Conant Ball in 1935. Russel Wright.

Armchair in mahogany with leather and ponyskin, designed by Russel Wright in 1932 for the boardroom of the Museum of Modern Art. Russel Wright.

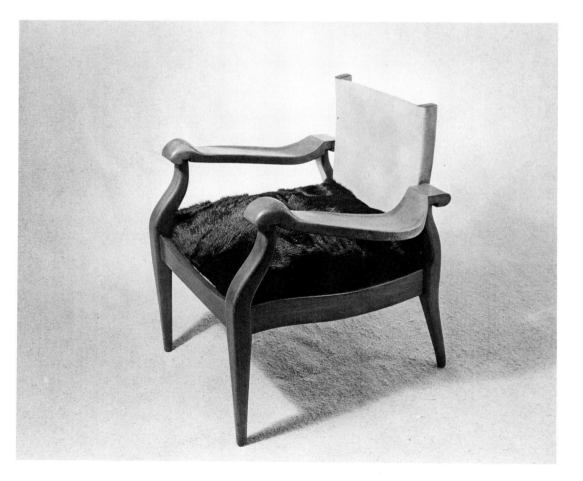

ambitious project called The American Way. Its objective was the development of a broad line of machine-made and hand-crafted products, in styles ranging from the sophisticated to the "country-made," that would promote design in industry as well as in craft production. It was a noble idea intended to stimulate an inherently American style of products that could be manufactured and sold at prices within the means of the average American family. Wright put everything he had into the project, writing royalty agreements with 72 manufacturers and sales contracts with 22 major stores across the country. The line of American Way products was launched in 1940 in an exhibition at Macy's department store. First Lady Eleanor Roosevelt gave the opening address, making note that the venture was a patriotic gesture as well as a cultural one. However, the dark clouds of war rising over Europe had already begun to divert attention from such ventures, and when the United States entered the war a year later the project was abandoned before it could gain practical headway. Its failure was the greatest disappointment of Wright's life, and he never regained his momentum.

Despite these examples of the broadening base of activities in design, it was difficult for American industrial designers to offset the opinion held by some that they were simply merchants of product aesthetics. After all, with few exceptions, the first generation of designers had come from the illusionary worlds of the theater, display, and advertising and had established a foothold in the economy by showmanship and dramatic success in showing industry how to sell its products through appearance. Technologists were obliged to accept designers as a necessary evil, and aestheticians reluctantly acknowledged their presumption to culture. Russell Lynes referred to the industrial designer as the best man at the wedding of art and industry, yet to him it was essentially a marriage of convenience. Only the public was fascinated by these new champions of its physical and cultural well-being. Only the public! Only the public had recognized that American industrial designers had found a form of expression that was all their own and of their own time.

Mary and Russel Wright. Courtesy of House and Garden; copyright 1933 (renewed 1961) by Condé Nast Publications, Inc.

The American Modern earthenware line, designed by Mary and Russel Wright in 1938, introduced the coupe shape and speckled glazes. Consumers took to the new line as a symbol of progress, and it became a runaway success. Russel Wright.

These products were displayed in the American Way exhibit at Macy's in 1940. The "American Way" was Russel Wright's dream of a consortium between industry and the best American artisans and industrial designers that would demonstrate that the United States had an indigenous design culture. The war destroyed that dream. American Magazine of Art, *November 1940.*

For better or worse, the marriage of art and industry had been consummated. While other countries were digging defenses, building armaments, and training men to kill, the Americans were blissfully transforming a swamp in Flushing and an island in San Francisco Bay into sites for twin world's fairs that were to be "a futile but magnificent gesture in the name of peace and prosperity." (61, 272)

In a virile decade the Americans had disassociated themselves from abject servility to European aesthetics. Their pride and their faith in the future were evident in the bold modernism of the New York fair's structures. Thousands walked the soaring ramp into the Perisphere to get a bird's-eye view of a world to come in the Democracity designed by Henry Dreyfuss. Then, they stood in line for hours to visit Norman Bel Geddes's General Motors exhibit, the Futurama, which had been expanded from his much-publicized Metropolis concept for the Shell Oil Company. As people emerged from Futurama they found themselves in a full-scale intersection of tomorrow for central Manhattan that is still to be realized. Industrial designers had a field day at the fair. Walter Dorwin Teague was not only one of the eight directors, but also designed exhibits for Eastman Kodak, Ford, and U.S. Steel. Other exhibits were designed by Donald Deskey, Russel Wright, Gilbert Rohde, George Sakier, and William Lescaze. Then, as the lights began to go out in one country after another, the American fairs began to seem like escapist exercises in futility. American designers, having established their value as agents of prosperity, began to redirect their interest—this time as servants of a nation in peril.

Just before Pearl Harbor brought this period to a close, the curtain of the future of American design was raised for a moment when the Museum of Modern Art announced the opening of twin competitions. The purpose of the first was to select a group of designers "capable of creating a useful and beautiful environment for today's living, in terms of furniture, fabrics and lighting." In addition, the museum agreed to make arrangements with manufacturers and merchandisers in order to make certain that the products selected would reach the typical American middle-income family. The objective of the second competition was to "discover designers of imagination and ability in the other Americas, and to bring . . . out suggestions on . . . the making of furniture for contemporary American requirements." (96, 27) The organizer of the competitions was Eliot Noyes (1910–1977), who had recently completed his studies in architecture under Walter Gropius at Harvard and was now director of the new department of industrial design at the museum. The jury, consisting of Alvar Aalto, Alfred H. Barr, Jr., Catherine H. Bauer, Edgar Kaufmann, Jr., and Edward D. Stone, awarded the principal prizes to Charles Eames (1907–1978) and Eero Saarinen (1910–1961) for a combination of chairs and storage pieces. Their solution for furniture was an upholstered shell of molded layers of veneer that echoed methods that were being developed for other mass-produced products, and their case furniture advanced the concepts of standardization, interchangeable parts, and functional adaptation. In September 1941 the products of the competition were exhibited at the museum under the title "Organic Design in Home Furnishings." Noyes defined *organic* as follows: ". . . harmonious organization of the parts within the whole, according to structure, material and purpose. Within this definition there can be no vain ornamentation or superfluity, but the part of beauty is none the less great—in ideal choice of material, in visual refinement, and in the rational elegance of things intended for use." (70, inside cover)

Edgar Kaufmann, Jr., credited Charles Eames as having attempted to carry the Bauhaus principles into deeper technology, but noted that his designs, like those of the Bauhaus, were to find their acceptance not in the average home but rather in contractual service. He also recognized the impact on the United States of Scandinavian design. "Nevertheless," Kaufmann wrote, "in the Thirties, it seemed to many of us that the period's design vitality was centered in America rather than in Europe, perhaps for the first time." (147, 142)

Air view of the New York
World's Fair of 1939–1940,
showing the theme struc-
tures, the Trylon and Perisphere conceived by
Wallace K. Harrison and
J. Andre Foyilhoux. In the
foreground are the Gen-
eral Motors and Ford
buildings. Architectural
Record, August 1940.

The Perisphere included
a miniature city—"Democ-
racity"—designed by
Henry Dreyfuss as a
"symbol of a perfectly
integrated metropolis
pulsing with life and
rhythm and music."
Architectural Record,
November 1938.

Futurama, a vision of 1960, was an extension of Norman Bel Geddes's Metropolis, the model "city of tomorrow" that was the center of a 1937 advertisement for the Shell Oil Company. It included expressways and an indoor sports arena. Hoblitzelle Theatre Arts Library, University of Texas, Austin.

The General Motors complex at the New York World's Fair (architect: Albert Kahn) included Bel Geddes's Futurama and a symbolic Manhattan intersection of tomorrow (shown at top). General Motors Corp.

Norman Bel Geddes in the Futurama's "Highways and Horizons" exhibit. The model of the Notre Dame cathedral was included for scale. The exhibit was intended to illustrate that "the world, far from being finished, is hardly begun yet." Hoblitzelle Theatre Arts Library, University of Texas, Austin.

Children in the Futurama. Viewers circled the exhibit in moving seats as though they were viewing it from the air. Cars moved along the expressways to lend realism. However, the smog (called traffic haze then, and taken as sign of progress) was missing. Hoblitzelle Theatre Arts Library, University of Texas, Austin.

*Charles Eames (left) and
Eero Saarinen in 1941,
when their furniture was
being selected for the
"Organic Design" award
at the Museum of Modern
Art. Organic Design in
Home Furnishings (MOMA
catalog).*

*Although conceived to
be built with laminated
plywood, Eames and Saa-
rinen's award-winning
chair proved too expen-
sive to produce. However,
after World War II, when
plastic made production
possible, this design
inspired many others.
Organic Design in Home
Furnishings (MOMA
catalog).*

Epilogue:
From Affluence to Conscience

In the end even this most prosperous and glamorous and complicated of professions comes down to a thing that is very old and very simple: one man's integrity against another's, one man's capacity as a working artist against another's, the vision with which he establishes his standards and the courage with which he sticks by them. The rest is trimming.

George Nelson, 1949 (159)

The second world war put an end for awhile to the design of competitive products and its promises for the future. Designers either joined the armed services or went into the factories to help produce material for war. Those few who remained in practice turned their energies to designing the products demanded by a nation in peril. Only rarely did a company assign or permit industrial designers to concern themselves with products aimed at postwar markets. As a result, designers were awakened to the part that they could play in products that are beyond open competition. They were called upon to apply their analytical capability and their sensitivity to form and utility in a climate that was controlled by technologists. Before the war they had been associated with and in many cases responsible to businessmen and merchandisers; now they were directed by scientists and engineers.

In the meantime, the man in the foxhole continued to dream about the wonderful world of tomorrow that designers had been promising him. Both he and the girl he had left behind saved for his return. Within one year after the end of the hostilities, 12 million American men were demobilized. Annual new-home starts went from 200,000 in 1945 to 1,154,000 in 1950. The volume of television sets manufactured went from zero in 1945 to 7,500,000 by 1950. Annual passenger-automobile production soared from 70,000 to 6,665,000 in the same period. The postwar boom was on. Every manufactured product had to be redesigned—first to shake off all vestiges of prewar aesthetic and wartime austerity, then to introduce the magic new materials, methodologies, and products that had been developed during the war, and finally to put into purchasable form the promises that had been made.

Design in the United States expanded from a comfortable practice for a few to big business for many. Degree programs in industrial design multiplied to more than twenty in order to meet the heavy demand for young talent. Design became an attractive profession for veterans, as well as for younger men and women completing high school, because of the opportunities it offered and the fact that it was associated with the exciting postwar environment. The spectacular demand for redesigned products convinced large and small manufacturers that design should become an integral component in corporate planning. Industrial design, therefore, which before the war had been offered primarily on a consulting basis, now became a part of management charged with bringing consistency in performance and appearance to all of a company's activities—not only the products themselves, but also packaging, display, graphics, signs, and facilities.

It was during this golden age that the concept of "good design" emerged. Museums and design and marketing centers offered their services to the general public as guides to a more efficient and aesthetically pleasing environment. Model homes were built and furnished in what the sponsors considered good taste. More often than not, however, the presentations were directed toward an exclusive audience of museumgoers, artists, designers, architects, and the like.

The United States government also got into the design business again, briefly, when it sought to secure peace and its world position by sharing its design experience by exporting teams of design specialists to assist industries on matters of marketable aesthetics in the Far East, South America, and the Arc of Asia. In addition, the United States contributed to the recovery of its former enemies by dispatching experts to assist them in their recovery and by opening up its schools and factories. Teams of German specialists toured the country, scores of Japanese students studied in our schools, and Americans conceived an exhibition ("Italy at Work") encouraging Americans to buy Italian goods again. The United States Information Agency staged exhibitions from Russia to Peru with the intention of stimulating the development of trade and production in other countries.

However, with prosperity and public acclaim dulling its judgment, design began to lose its sense of balance and public responsibility. The consumer, caught in a frenzy of postwar marketing, became something to sell things to rather than someone to serve. Product obsolescence for the sake of sales found designers being asked, or even recommending, that irrelevant form manipulation and irreverent ornamentation be used to artificially stimulate the market. Conventional wisdom had it that the public must be given what it wanted and that it could be guided to what it wanted by mass-media persuasion. In 1956, Chrysler—the last bastion of rationalism in design—exclaimed "Suddenly it's 1960" and exploded in a fit of fins and extraneous embellishment. American products became caricatures of their function and, with the new architecture, reflected the carnival-like consumption of the immediate postwar period. The golden age was proving to be an orgy.

It is difficult to determine exactly when American consumers began to be disenchanted with the quality of their environment and the character and the quality of the products that were being touted by the media and thrust at them by superstores and overflowing discount houses. Was it when Volkswagens were first brought into the country? Was it when the first young family decided on a backpacking vacation rather than a stay at a motel with Olympic pool and Tudor drinking den? Did Rachel Carson set it off with *Silent Spring*, or Ralph Nader with *Unsafe At Any Speed*? When did ecology and pollution become part of everyday conversation? And when did industry and the government begin to allow that humanity was tolerated on this planet only by a fragile biosphere? Should one credit Sputnik for proving what everyone really knew already—that earth was, indeed, only a small spaceship adrift in the firmament?

The problem was first admitted publicly in 1962 when President John F. Kennedy found it necessary to call attention to the abuse of the con-

sumer by proclaiming that his inalienable rights in a technological world also included the right to safety, the right to be informed, the right to choose, and the right to be heard. Kennedy's recommendation that national measures be established by which the quality of the man-made environment could be monitored in the public interest is slowly being transformed into actions to serve the public welfare. This may be the first time in history when, in contrast to Emerson's warning that "things are in the saddle and ride mankind," things are being brought to account by man. As a result, by a series of slow and often painful steps a profound change has been underway in the United States. The callous impersonality of the product is being tempered by the quality of responsibility that is being mandated by public sentiment. A new morality is emerging by which the tyranny of a gross national rate of production may be displaced by the democracy of a volume and quality of production that is directly matched to the needs of the people.

Bibliography

1 Adams, Charles Francis. *Familiar Letters of John Adams and His Wife, Abigail Adams, During the Revolution*. New York: Hurd and Houghton, 1876.

2 Allen, Lewis Falley. *Rural Architecture*. New York: Orange Judd, 1852.

Andrews, Edward D. and Faith. *Shaker Furniture: The Craftsmanship of an American Communal Sect*. 1937. New York: Dover, 1964.

3 Andrews, Wayne. *Architecture, Ambition and Americans*. New York: Harper and Row, 1964.

Appelbaum, Stanley. *The Chicago World's Fair of 1893*. New York: Dover, 1980.

4 Bancroft, Hubert Howe. *The Book of the Fair*, volume I. New York: Bounty, 1894.

5 Banham, Reyner. *Theory and Design in the First Machine Age*. New York: Praeger, 1967; Cambridge: MIT Press, 1980.

6 Bathe, Greville and Dorothy. *Oliver Evans: A Chronicle of Early American Engineering*. Philadelphia: Historical Society of Pennsylvania, 1935.

Battersby, Martin. *The Decorative Twenties*. New York: Walker, 1969.

7 Baxter, Richard. *Christian Directory*. London, 1678.

8 Bayer, Herbert, and Gropius, Walter and Ise, editors. *Bauhaus*. New York: Museum of Modern Art, 1938.

9 Beecher, Catherine E., and Stowe, Harriet Beecher. *The American Woman's Home*. New York: J. B. Ford, 1869.

10 Beecher, Catherine. *A Treatise on Domestic Economy*. New York: Harper and Brothers, 1849.

Bel Geddes, Norman. See refs. 37 and 140.

Bennett, Charles Alpheus. *History of Manual and Industrial Education, 1870–1917*. Peoria: Charles A. Bennett Company, 1937.

Berman, Eleanor Davidson. *Thomas Jefferson Among the Arts*. New York: Philosophical Library, 1947.

11 Bishop, John Leander. *A History of American Manufactures, 1608–1868*. New York: Johnson Reprint Corporation, 1967.

12 Bogardus, James. *Cast Iron Buildings, Their Construction and Advantage*. New York: J. W. Harrison, 1856.

Bode, Carl. *American Life in the 1840s*. New York University Press, 1967.

Bok, Edward. *The Americanization of Edward Bok*. New York: Charles Scribner's Sons, 1923.

Bolles, Albert S. *Industrial History of the United States*. Norwich: Henry Bill, 1879.

Bridenbaugh, Carl. *The Colonial Craftsman*. University of Chicago Press, 1966.

Briggs, Asa, editor. *William Morris: Selected Writings and Designs*. Baltimore: Penguin, 1962.

13 Buckles, Robert A. *Ideas, Inventions and Patents*. New York: Wiley, 1957.

Burchard, John, and Bush-Brown, Albert. *The Architecture of America*. Boston: Little, Brown, 1961.

Bush, Donald J. *The Streamlined Decade*. New York: George Braziller, 1975.

Butterfield, Roger. *The American Past*. New York: Simon and Schuster, 1966.

Byrn, Edward W. *The Progress of Invention in the Nineteenth Century*. New York: Munn, 1900.

14 Calkins, Ernest Elmo. *And Hearing Not*. New York: Charles Scribner's Sons, 1946.

15 Carlyle, Thomas. "Sign of the Times," in *Critical and Miscellaneous Essays*. Boston: Dana Estes, 1869.

16 Carpinael, William. *Registration of Design*. London: Alexander McIntosh, 1851.

17 Carter, Clarence E., editor. *The Correspondence of General Thomas Gage*. New Haven: Yale University Press, 1933.

18 Cathers, David M. *Furniture of the American Arts and Crafts Movement*. New York: New American Library, 1981.

19 Cheney, Sheldon and Martha. *Art and the Machine*. New York: Whittlesey House, McGraw-Hill, 1936.

20 Chevalier, Michel. *Society, Manners and Politics in the United States*. Boston: Weeks, Jordan, 1839.

Cigrand, B. J. *Story of the Great Seal of the United States*. Chicago: Cameron, Amberg, 1903.

Clark, Isaac A. *Art and Industrial Education (1876–1904)*. Philadelphia: U. S. Bureau of Education.

Clark, Robert Judson. *The Arts & Crafts Movement in America, 1876–1916*. Princeton University Press, 1972.

Clark, Victor S. *History of Manufactures in the United States, 1607–1860,* volume I. New York: McGraw-Hill, 1929.

Clarke, Hermann Frederick. *John Coney.* Cambridge: Houghton Mifflin/Riverside Press, 1932.

21 Coles, William A. *Architecture and Society, Selected Essays of Henry van Brunt.* Cambridge: Belknap Press of Harvard University Press, 1969.

Condit, Carl W. *American Building Art: The Nineteenth Century.* New York: Oxford University Press, 1960.

22 Cooper, James Fenimore. *Notions of the Americans: picked up by a Traveling Bachelor.* Philadelphia: Carey, Lea and Carey, 1828.

23 Coxe, Tench. *An Enquiry into the Principles on Which a Commercial System for the United States Should be Founded* (Commercial Pamphlet Volume 3, Number 6). Philadelphia: Library of Congress, 1787.

Coxe, Tench. *A View of the United States of America in a Series of Papers Written at Various Times in the Years Between 1787 and 1794.* New York: Augustus M. Kelley, 1965.

24 Craig, Lois. *The Federal Presence.* Cambridge: MIT Press, 1978.

25 Crevecoeur, Michel-Guillaume Jean de. *Letters from an American Farmer.* London, 1792; New York: Dutton, 1957.

26 Davidson, Marshall. *Life in America,* volume I. Boston: Houghton Mifflin, 1951.

Davy, John, editor. *The Collected Works of Sir Humphry Davy.* London: Smith, Elder, 1839–40.

de Tocqueville, Alexis. See ref. 86.

De Zurko, Edward Robert. *Origins of Functionalist Theory.* New York: Columbia University Press, 1957.

27 Dreyfuss, Henry. *Designing for People.* New York: Paragraphic, 1967.

28 Dunlap, William. *History of the Rise and Progress of the Arts of Design in the United States.* New York: Scott, 1834.

29 Eastlake, Charles L. *Hints on Household Taste in Furniture, Upholstery and Other Details.* New York: Dover, 1969 (reprint).

30 Edwards, J. E. *A Complete History or Survey of all the Dispensation and Methods of Religion.* London, 1694.

Emmet, Boris, and Jeuck, John E. *Catalogues and Counters.* University of Chicago Press, 1950.

Fales, Martha Gandy. *Early American Silver for the Cautious Collector.* New York: Funk and Wagnalls, 1970.

31 Fiske, John. *The Beginnings of New England.* Boston: Houghton Mifflin, 1902.

32 Foner, Philip S. *Basic Writings of Thomas Jefferson.* New York: Wiley, 1944.

33 Ford, Henry. *My Life and Work.* New York: Doubleday, 1922.

34 Ford, Paul Leicester, editor. *Thomas Jefferson, Notes on the State of Virginia.* New York: Putnam, 1892.

35 Frankl, Paul. *Form and Re-Form.* New York: Harper and Brothers, 1930.

Frankl, Paul. *New Dimensions.* New York: Payson and Clarke, 1928.

36 Franklin, William Temple, editor. *Private Correspondence of Benjamin Franklin,* volume II. London: Henry Colburn, 1833.

Gayle, Margot, and Gillon, Edmund V., Jr. *Cast-Iron Architecture in New York.* New York: Dover, 1974.

37 Bel Geddes, Norman. *Horizons.* Boston: Little, Brown, 1932.

Giedion, Siegfried. *Mechanization Takes Command.* New York: Oxford, 1948.

38 Gloag, John. *A Social History of Furniture Design.* London: Cassel, 1966.

39 Greeley, Horace. *Art and Industry as represented in the Exhibition at the Crystal Palace, New York, 1853–4.* New York: Redfield, 1853.

40 Greenough, Horatio. *Form and Function.* Berkeley: University of California Press, 1947.

41 Greenough, Horatio. *The Travels, Observations and Experiences of a Yankee Stonecutter.* Gainsville: Scholar's Facsimiles and Reprints, 1958.

42 Griffiths, John W. *Treatise on Marine and Naval Architecture.* London: George Philips and Son, 1856.

43 Harper, Robert. Jo Sinel, Father of American Design (tape transcript). Meyer Library, California College of Arts and Crafts, Oakland, 1972.

Hawkins, Layton S. *Development of Vocational Education.* Chicago: American Technical Society, 1951.

44 Hobson, Charles F., and Rutland, Robert. *The Papers of James Madison.* Charlottesville: University Press of Virginia, 1981.

45 Hornung, Clarence P. *Treasury of American Design.* New York: Abrams, 1950.

Hornung, Clarence P. *Wheels Across America.* New York: Barnes, 1959.

46 Hutcheson, Harold. *Tench Coxe: A Study in American Economic Development.* Baltimore: Johns Hopkins Press, 1938.

47 Israel, Fred L., editor. *The State of the Union Messages of the Presidents.* New York: Chelsea House, Robert Hector, 1966.

48 Jarves, James Jackson. *The Art-Idea* (1864). Cambridge: Belknap Press of Harvard University Press, 1960.

49 Jarves, James Jackson. *Art Thoughts.* New York: Hurd and Houghton, 1869.

50 Jeremy, David L. *Transatlantic Industrial Revolution: The Diffusion of Textile Technologies Between Britain and America, 1790–1830s.* Cambridge: MIT Press, 1981.

51 Jefferson, Thomas. *The Papers of Thomas Jefferson.* (Julian P. Boyd, editor.) Princeton University Press, 1953.

52 Jefferson, Thomas. *The Writings of Thomas Jefferson.* (H. Washington, editor.) Washington: Taylor and Maury, 1853.

Johnson, Philip. *Machine Art.* New York: Museum of Modern Art/Norton, 1934.

Kasson, John F. *Civilizing the Machine.* New York: Penguin, 1977.

53 Keppel, Frederick P., with Duffus, R. L. *The Arts in American Life.* New York: McGraw-Hill, 1933.

Kimball, Sidney Fiske. *Mr. Samuel McIntire Carver, the Architect of Salem.* Portland: Southworth Anthaensen Press for Essex Institute of Salem, Massachusetts, 1940.

54 Klemm, Friedrich. *A History of Western Technology.* Cambridge: MIT Press, 1964.

Kouwenhoven, John A. *Adventures of America, 1857–1900.* New York: Harper, 1938.

55 Kouwenhoven, John Atlee. *Made in America.* New York: Doubleday, 1948.

Krause, Joseph H. *The Modern Design Concept: Origin and Development.* University of Southern California, 1963.

56 Labaree, Leonard Woods. *The Papers of Benjamin Franklin,* volume 2. New Haven: Yale University Press, 1960.

57 Laslett, Peter. *John Locke, Two Treatises of Government.* Cambridge University Press, 1967.

58 Latrobe, Benjamin Henry, papers of (microtext for Maryland Historical Society). Baltimore: J. White, 1976.

Lea, Zilla Rider. *The Ornamental Chair: Its Development in America.* Rutland: Tuttle, 1960.

59 Leslie, Frank. *Illustrated Historical Register of the Centennial Exposition, 1876.* New York: Paddington, 1974.

Lifshey, Earl. *The Housewares Story.* Chicago: National Housewares Manufacturers Association, 1973.

60 Loewy, Raymond. *Never Leave Well Enough Alone.* New York: Simon and Schuster, 1951.

Lynes, Russell. *The Domesticated Americans.* New York: Harper and Row, 1963.

61 Lynes, Russell. *The Taste Makers.* New York: Harper, 1954.

MacCarthy, Fiona. *All Things Bright and Beautiful.* University of Toronto Press, 1972.

Mackay, James. *Turn-of-the-Century Antiques: An Encyclopedia.* New York: Dutton, 1974.

62 Malone, Dumes, editor. *Correspondence between Thomas Jefferson and Pierre Samuel du Pont de Nemours, 1798–1817.* Boston: Houghton Mifflin, 1930.

63 Marx, Leo. *The Machine in the Garden.* New York: Oxford University Press, 1967.

64 Metropolitan Museum of Art. *19th-Century America: Furniture and Other Decorative Arts.* New York: New York Graphic Society, 1970.

Metzger, Charles. *Emerson and Greenough: Transcendental Pioneers of an American Aesthetic.* Berkeley: University of California Press, 1954.

65 Moholy-Nagy, Laszlo. *Vision in Motion.* Chicago: P. Theobald, 1947.

66 Morris, William. *Signs of Change.* London: Longmans, Greene, 1903.

67 Mowry, George, editor. *The Twenties: Fords, Flappers and Fanatics.* Englewood Cliffs: Prentice-Hall, 1963.

Mumford, Lewis. *Art and Technics.* New York: Columbia University Press, 1952.

68 Mumford, Lewis. *The Culture of Cities.* New York: Harcourt, Brace, 1938.

Mumford, Lewis, editor. *Roots of Contemporary American Architecture.* New York: Dover, 1972.

69 Munz, Ludwig, and Kunstler, Gustav. *Adolf Loos, Pioneer of Modern Architecture.* New York: Praeger, 1966.

Naylor, Gillian. *The Bauhaus.* London: Studio Vista, 1968.

Nevins, Allan, and Hill, Frank Ernest. *Ford: Expansion and Challenge: 1915-1933.* New York: Scribner, 1956.

70 Noyes, Eliot. *Organic Design in Home Furnishings.* New York: Museum of Modern Art, 1941.

Otto, Celia Jackson. *American Furniture of the 19th Century.* New York: Viking, 1965.

71 Phillips, John Marshall. *American Silver.* New York: Chanticleer, 1949.

72 Preble, George Henry. *Origins and History of the American Flag.* Philadelphia: Brown, 1917.

Reed, Robert. *The Streamline Era.* San Marino: Golden West, 1975.

73 *Report of Commission to International Exposition of Modern Decorative and Industrial Art in Paris, 1925.* Washington: U.S. Department of Commerce, 1925.

74 Richards, Charles R. *Art in Industry.* New York: National Society for Vocational Education and Department of Education of the State of New York, 1922.

Rodest, Howard B. *Toward Common Ground: The Story of Ethical Societies in the United States.* New York: Unger, 1969.

75 Rodgers, Charles T. *American Superiority at the World's Fair.* Philadelphia: Hawkins, 1852.

Roe, Joseph W. *English and American Tool Builders.* New Haven: Yale University Press, 1916.

76 Rosenberg, Nathan. *The American System of Manufacturing.* Edinburgh University Press, 1969.

77 Rourke, Constance Mayfield. *The Roots of American Culture and Other Essays.* New York: Harcourt, Brace, 1942.

Rowsome, Frank Jr. *Trolley Car Treasury.* New York: Bonanza, 1956.

78 Schaefer, Herwin. *Nineteenth Century Modern: The Functional Tradition in Victorian Design.* New York: Praeger, 1970.

79 Schmutzler, Robert. *Art Nouveau.* New York: Abrams, 1962.

80 Siegfried, Robert. *Humphrey Davy on Geology: The 1805 Lectures for the General Audience.* Madison: University of Wisconsin Press, 1980.

Sloane, Eric. *A Museum of Early American Tools.* New York: Wilfred Funk, 1964.

81 Smith, Adam. *An Inquiry into the Natures and Causes of the Wealth of Nations* (1778). New York: Modern Library/Random House, 1937.

Smith, Henry Nash, editor. *Popular Culture and Industrialism.* New York University Press, 1967.

Smith, Walter. *Art Education.* Boston: Osgood, 1872.

82 Smith, Walter. *The Masterpieces of the Centennial International Exhibition,* volume II. Philadelphia: Gebbie and Barrie, 1876.

The Story of the United States Patent Office, 1790-1956. Washington: U.S. Government Printing Office, 1956.

Streichler, Jerry. *The Consultant Industrial Designer in American Industry from 1927-1960.* Ann Arbor: University Microfilms, 1963.

83 Sullivan, Louis H. *The Autobiography of an Idea.* New York: Dover, 1956.

84 Taussig, F. W. *State Papers and Speeches on the Tariffs.* Cambridge: Harvard University Press, 1893.

Teague, Walter Dorwin. *Design this Day.* New York: Harcourt, Brace, 1940.

85 Thorn, C. Jordon. *The Handbook of American Silver and Pewter Marks.* New York: Tudor, 1949.

86 Tocqueville, Alexis de. *Democracy in America.* New York: Schocken, 1961.

87 Toynbee, Arnold. *The Industrial Revolution* (1884). Boston: Beacon, 1960.

True, Webster Prentiss. *The Smithsonian: America's Treasure House.* New York: Sheridan House, 1950.

Tryon, Rolla M. *Household Manufactures in the United States 1640-1860.* University of Chicago Press, 1917.

Usher, Abbott P. *The History of Mechanical Inventions.* New York: McGraw-Hill, 1929.

Veblen, Thorstein. *The Theory of the Leisure Class.* New York: Viking, 1934.

Veronesi, Julia. *Style and Design 1909–1929.* New York: George Braziller, 1968.

Wallace, Don. *Shaping America's Products.* New York: Reinhold, 1956.

88 Washington, H. A., editor. *The Writings of Thomas Jefferson.* Washington: Taylor and Maury, 1853.

89 Watkinson, Ray. *William Morris as Designer.* New York: Reinhold, 1967.

90 Willcocks, Lewis. *Report on a Plan for Extending and More Perfectly Establishing the Mechanical and Scientific Institutions of New York.* New York: Daniel Fanshaw, 1824.

Williams, Raymond. *Culture and Society, 1780–1950.* London: Chatto and Windus, 1958.

Wilson, Joseph M. *The Masterpieces of the Centennial International Exhibition,* volume III. Philadelphia: Gebbie and Barrie, 1876.

Wilson, Mitchell. *American Science and Invention.* New York: Simon and Schuster, 1954.

Wright, Louis B., et al. *The Arts in America: The Colonial Period.* New York: Charles Scribner's Sons, 1966.

Wright, Sidney C. *The Story of the Franklin Institute.* Philadelphia: Franklin Institute, 1938.

91 Wright, Frank Lloyd. *An Autobiography.* New York: Horizon, 1977.

92 *American Magazine of Art,* June 1918. "Resolutions Adopted by the American Federation of Arts in Convention."

93 *Architectural Forum,* May 1934. "Art and Machines."

94 *Architectural Forum,* December 1934. "Contemporary Quinquennial."

95 *Architectural Forum,* August 1940.

96 *Architectural Forum,* October 1940. "Design Decade."

Arts and Decoration (refs. 97–113)

97 Purdy, Frank. "Some Facts about Industrial Art." May 1920.

98 Purdy, Frank. "Art in American Industry." August 1920.

99 "Forecast of November Automobile Salon." November 1920.

100 Purdy, Frank. "The Taste of the American People." November 1920.

101 Purdy, Frank. "The New Museum." December 1920.

102 Purdy, Frank. "America Needs Co-operation in Industrial Art." January 1921.

103 Bach, Richard F. "A Note of Progress in Industrial Art." February 1921.

104 "Poiret, Interpreter of His Own Age." March 1921.

105 Levy, Florence. "The Hope of the World in Art." May 1921.

106 Price, Matlack. "The Hand of the Designer in Advertising Art." July 1921.

107 Price, Matlack. "Practicality, Imagination and the Designer." July 1921.

108 Haraucourt, Edmond. "A Salon of French Taste." December 1921.

109 Randole, Leo. "Art Wedded to Industry." September 1922.

110 Ackerman, Phyllis. "The Artist Decorators of France Have a Salon." August 1923.

111 McCann, E. Armitage. "A Significant Showing of Industrial Art." October 1923.

112 "Fine Exhibition of Beautiful Interiors." November 1924.

113 "Beautiful Antique and Modern Home Furnishings." February 1925.

114 *Atlantic Monthly,* November 1872. "Walter Smith's Art Education."

115 Bach, Richard. "Production Calls Art and Color to its Aid." *Manufacturing Industries,* December 1928.

116 *Boston Evening Post,* January 18, 1773.

117 Brooker, Bertram R. "Beauty's Place in Business." *Printer's Ink,* April 5, 1928.

118 *Business Week,* January 29, 1930. "The Eyes Have It."

119 *Business Week,* October 28, 1931. "1931 Industrial Art Show Eliminates Curlicues of 1925."

120 Calkins, Ernest Elmo. "Advertising Art in the United States." *Modern Publicity,* 1930, volume 7.

121 Calkins, Ernest Elmo. "Beauty, the New Business Tool." *Atlantic Monthly,* August 1927.

122 *Columbian Magazine,* September 1786.

123 Commager, Henry Steele. "The Search for a Usable Past." *American Heritage,* February 1965.

124 Cortissoz, Royal. "A Contemporary Movement in American Design." *Scribner's Magazine,* May 1929.

125 *The Craftsman,* October 1901.

126 Dreyfuss, Henry. "Everyday Beauty." *House Beautiful,* November 1933.

127 Fitch, James Marston. "When Housekeeping Became a Science." *American Heritage,* August 1961.

128 *Forbes,* April 1, 1934. "Best Dressed Products Sell Best."

129 *Fortune,* February 1934. "Both Fish and Fowl."

The Franklin Journal and American Mechanics Magazine (refs. 130–132)

130 January 1826.

131 February 1826.

132 September 1826.

The Journal of the Franklin Institute (refs. 133–138)

133 June 1829.

134 August 1838.

135 January 1840.

136 April 1842.

137 November 1842.

138 January 1844.

139 Freund, F. E. Washburn. "The Lesson of German Applied Art." *International Studio,* July 1922.

140 Bel Geddes, Norman. "Ten Years from Now." *Ladies' Home Journal,* January 1931.

Harper's Monthly Magazine (refs. 141–143)

141 Knight, Edward H. "The First Century of the Republic." February 1875.

142 Wells, David A. "The First Century of the Republic." April 1875.

143 "Fret-sawing and Wood-Carving." March 1878.

144 Advertisement in *House and Garden,* November 1933.

Jensen, Oliver. "Farewell to Steam." *American Heritage,* December 1957.

145 *Journal of the Design and Industries Association,* number 11, Summer 1919.

146 Kaufmann, Edgar. "Borax, the Chromium-Plated Calf." *Architectural Review,* August 1948.

147 Kaufman, Edgar. "Viewpoints." *Interiors,* September 1972.

148 Kurtzworth, H. M. "Industrial Art, A National Asset." *Industrial Education Circular* number 3, May 1919.

149 *Literary Digest,* April 26, 1924. "Machine-Made Beauty."

150 London *Times,* September 2, 1851.

151 Ludlow, Fitz Hugh. "The American Metropolis." *Atlantic Monthly,* January 1865.

152 Lyon, Peter. "Isaac Singer and his Wonderful Sewing Machine." *American Heritage,* October 1958.

153 *The Metropolitan Museum of Art Bulletin,* December 1930.

154 *The Metropolitan Museum of Art Bulletin,* October 1931.

155 Mumford, Lewis. "American Taste." *Harper's Monthly,* October 1927.

156 Mumford, Lewis. "Beauty and the Industrial Beast." *New Republic,* June 6, 1923.

157 Mumford, Lewis. "Machinery and the Modern Style." *New Republic,* August 31, 1921.

158 Mumford, Lewis. "Machines for Living." *Fortune,* February 1933.

159 Nelson, George. *Design,* November 1949.

New York Times (refs. 160–191)

160 February 4, 1900, 23:2.

161 April 28, 1900, 8:3.

162 July 29, 1900, 16:3.

163 September 18, 1900, 11:6.

164 June 8, 1902, 8:1.

165 November 8, 1903, 6:4.

166 January 21, 1904, 8:4.

167 May 12, 1913, 8:7.

168 September 21, 1913, VII 16:4.

169 January 4, 1914, VII 9:3.

170 January 11, 1914, II 10:5.

171 August 16, 1914, VI 8:3.

172 June 10, 1917, VI 14:1.

173 February 14, 1918, 10:7.

174 June 2, 1918, VII 15:3.

175 June 9, 1918, VI 14:1.

176 December 22, 1918, IV 8:2.

177 February 11, 1919, 10:7.

178 October 26, 1919, IV 10:2.

179 January 11, 1920, IX 3:4.

180 April 18, 1920, VII 4:1.

181 May 9, 1920, VI 9:1.

182 October 31, 1920, III 20:1.

183 January 2, 1921, III 20:3.

184 October 2, 1921, III 20:1.

185 February 26, 1922, II, p. 4.

186 April 16, 1922, VII 8:4.

187 January 31, 1926, IV 2:2.

188 January 6, 1929, VIII 4:5.

189 January 27, 1929, IX 4:5.

190 October 14, 1931, 24:4.

191 April 16, 1935, 23:1.

192 "Effects of Machinery." *North American Review,* January 1832.

193 Olsen, Alma Louise. "Modern Swedish Decorative Art." *Art and Archaeology,* February 1927.

194 Richards, Charles. "Design for the Craftsman and Design for the Machine." *House Beautiful,* August 1931.

195 Richards, Charles. "The Third International Exhibition of Industrial Art." *American Magazine of Art,* November 1930.

196 Sargent, Walter. "The Training of Designers." *American Magazine of Art,* September 1918.

197 Seitlin, Percy. *PM Magazine,* August–September 1938.

Seldes, Gilbert. "Artist in a Factory." *New Yorker,* August 29, 1931.

198 Shidle, N. "Beauty Doctors Take a Hand in Automotive Design." *Automotive Industry,* August 13, 1927.

199 Solon, Leon V. "The Fostering of American Industrial Art by the Metropolitan Museum." *Architectural Record,* February 1927.

200 Tiffany, Louis C. "The Gospel of Good Taste." *Country Life in America,* November (mid-month) 1910.

201 Walker, Timothy. "Defense of Mechanical Philosophy." *North American Review,* July 1831.

202 Wells, Arnold. "Father of our Factory System." *American Heritage,* April 1958.

Whitridge, Arnold. "Eli Whitney: Nemesis of the South." *American Heritage,* April 1955.

203 Young, Grace Alexandra. "Designers at Work in America." *PM Magazine,* August–September 1938.

204 Bach, Richard. "The Architect and the Industrial Arts—An Exhibition of Contemporary American Design" (catalog). Metropolitan Museum of Art, February 1929.

205 Blake, William P. "Sketch of the Life of Eli Whitney, the Inventor of the Cotton Gin." *New Haven Colony Historical Papers,* November 28, 1887.

206 Chapman, Dave. Letter to author, July 29, 1963.

207 Davis, Gage. Letter to author, September 9, 1965.

208 *Encyclopedia of World Art,* 1968, s.v. "Industrial Design," volume 8, column 100 (author: Gillo Dorfles).

209 *Encyclopedia Americana,* 1956, volume 12.

210 *Encyclopedia Americana,* 1957, volume 29.

211 *General Management Series Number 99.* American Management Association, 1929.

212 Gropius, Walter. Lecture to Birmingham Design and Industry Association, March 15, 1935.

213 Guild, Lurelle. Letter to author, July 9, 1963.

214 Hoover Institute Archives, Box 7 (New York City, September 2, 1925). Stanford University.

215 Kimball, Abbott. "Dollar and Cents Value of Beauty" (pamphlet). New York: Industrial Institute of the Art Center, 1930s.

216 "Selected Collection of Objects from the International Exposition of Modern Decorative and Industrial Art" (catalogue of exhibit). American Association of Museums, 1925.

217 Chapman, Dave. Letter to author, July 29, 1963.

218 Sandin, Ray. Letter to author, June 1963.

219 Vassos, John. Letter to author, July 26, 1976.

220 Wright, Russel. Interview by author.

Index

American Design Ethic

The MIT Press
Cambridge, Massachusetts
London, England